CARLSBAD, CAVES, AND A
ROBERT NYMEYER
CAMERA

CAVE BOOKS   DAYTON, OHIO

CAVE BOOKS
4700 Amberwood Drive
Dayton, OH 45424
www.cavebooks.com
Publications Affiliate of the Cave Research Foundation

Library of Congress Cataloging in Publication Data

Nymeyer, Robert, 1910–1983.
    Carlsbad, caves, and a camera.

    Rerint of the 1978 ed. published by Zephyrus Press, Teaneck, NJ.

    Includes index.

    1. Caves—New Mexico—Carlsbad region. 2. Carlsbad
region, N.M.—Description and travel. 3. Nature photography.
I. Title.
GB605.N3N95 1993   917.89'42   78-6550
ISBN 13: 978-939748-36-5 pbk.
ISBN 10: 0-939748-36-3 pbk.

Contents

To Dave and Ted and Glenn and Charlie and "Static" and Seth, and in memory of Tommie and Julian and Sonnie and B. A., for without them these events could never have taken place

CARLSBAD, CAVES, AND A CAMERA

Prologue

THE LOVE OF CAVING must have been born in me. I was intrigued at an early age by a collection of old photographs from some of my mother's albums. These were taken about the turn of the century in one of the first caves in the Guadalupes known to have been visited by white men; it was named McKittrick Cave for a pioneer rancher who settled here about 1880. No one knows if he ever entered the cave which bears his name, but it became a favorite picnic spot for residents of the little village of Eddy, which later was renamed Carlsbad. Mother's photographs show the cave before vandals started wrecking it, the custom of the time being to bring out an armload of "pretty formations" as souvenirs of a fun-filled Sunday. The pictures also show the early spelunkers' mode of dress for the occasion, and today's cavers will stare in disbelief.

The first cave I ever entered was the queen of them all, Carlsbad Caverns, when I was sixteen. The immensity of this cave is so overwhelming, however, that the cave bug did not get a chance to bite, although I had become a prime target. The irresistible lure of caving actually hit me in a small cave a few miles north of Carlsbad Caverns. This was in 1933 in little Spider Cave when I was twenty-three years old. For the first time I crawled up an unexplored tunnel, negotiated a tight pinch, and stepped out into a virgin cave stretching far beyond the limits of my flashlight. For the first time I knew the thrill of "She goes!"

The quest for that thrill led me into as many caves as we could find in the decade that followed. Speleology was an unknown word, caves were generally held in disrepute as potential danger spots, the decorations in them were simply "formations," and people who crawled about in them for fun were undoubtedly a little crazy. I guess we were indeed a little balmy—cavers today would judge us so. Hard hats were worn only by miners; we wore sombreros, worn-out fedoras, English-billed caps, or just knitted skull caps, anything that would absorb a little of the shock if we raised up under a low pitch or a firm stalactite. Pants and shirts were anything of material heavy enough to stand the strain; footwear was whatever might be lying about of sufficient age to be worth little else. We did not know the difference between a Prusik knot and a bowline, and "belay" was a nautical expression. It is unbelievable that we got out of some of the messes we found ourselves in with as little trouble as we did; the fact that we escaped serious injury or even death remains almost a miracle.

My first attempt at cave photography was made with only light coming through an entrance, so the field was strictly limited. Wanting to photograph the dark areas led me to try time exposures by lantern light. Some of these were fairly successful, but they only whetted a desire for something better. I tried flashpowder for the first time in Black Cave, and the results were disappointing, for I had yet to learn to keep the flare of the flash out of the lens, but the lesson learned there led to better results. The process I finally hit on was simple, extremely efficient, and gadget-free. It consisted of a couple of dozen pages from a pulp magazine—*Western Story* was always dependable— a 16-ounce jar of flashpowder, and a handful of matches.

After choosing the site for a photograph, I would roll up a magazine page into a firm bar, leaving one end cupped, like a spoon. Placing this out of camera range against a white wall or stalagmite, which served as a reflector, I would pour a charge of powder into the "spoon." The amount was determined by the size of the scene being shot, ranging from a thimbleful to several ounces. After opening the camera on time exposure, I would light the "fuse" end of the page, and, if I desired, run around to get in the picture before the paper burned to the powder and ignited it. The resulting flash presented one problem we never solved: Only one photograph could be made in a room or section, because the smoke resulting from the burning powder thoroughly fogged up the scenery for an hour or so.

From the single flash picture, the next natural step led to multiple-flash photographs when we found an extremely large vista for which we knew one flash would not do the job. It was a simple matter to place two or more "spoons" of flashpowder at appropriate sites, station a cave explorer at each, and then have all light the "fuses" in unison upon notice from the photographer standing by the camera. The only precaution was to avoid shining flashlights in the direction of the camera after the shutter had been opened. Surprisingly, because the pulp tended to burn evenly, we seldom had one flash precede the others so much that its smoke intruded too much into the picture. This was synchronized flash at its most primitive.

We soon found flashpowder too expensive for our strained wallets —the Great Depression was a painful reality, and although products were cheap, money was scarce. I discovered that I could buy magnesium powder at bulk rate for a fraction of what flashpowder cost. But we also discovered that the blaze from burning pulp paper simply was not hot enough to ignite the slower burning magnesium. Cavers, however, are nothing if not ingenious. We experimented and found that a little shotgun powder mixed with the magnesium would produce enough heat to set it off. So we started taking only the more certain shots at our dove, quail, and ducks, and sacrificed a few precious 12-gauge shells for powder.

We never used flashbulbs. For one thing, they were just coming onto the market and were big and bundlesome, about the size of today's 300-watt light bulbs, easily shattered in a tight cave squeeze. An eight-fluid ounce medicine bottle of flashpowder would light as many photographs as two dozen of the things. And, second, they were just too darned expensive.

Cave photographs in color? Impossible! Color film for the cave explorer was still a dream of the future.

After my first two or three attempts at cave photography with a cheap folding pocket camera of uncertain origin and poor quality lens, I realized that if my underground photographs were to improve, I would have to get a better camera. I managed to trade for an excellent used folding Eastman 2-C Autographic Kodak Jr. This used the size 130 roll film with six exposures, making a negative $2\frac{7}{8}$ by $4\frac{7}{8}$ inches, only slightly smaller than the popular postcard cameras of that day.

The camera was equipped with an Ilex shutter with time and bulb exposure, and timed speeds from one second to 1/100th of a second. The lens was an Eastman Anastigmat F:7.7, not very fast when compared with today's high-speed jobs, but it made negatives of fine definition and detail. It focused by a sliding front from infinity to six feet, and I modified this somewhat so as to be able to take sharp pictures down to a distance of only three feet.

I had the local saddle maker fashion a holster case of heavy leather with sufficient depth to carry two rolls of film. Camera, case, and film comprised a compact package that weighed only about three pounds and offered little hindrance in crawling through the caves.

After one attempt at carrying a cumbersome tripod through a cave, I decided it was not worth the effort. The tripod constantly got in the way, its dangling legs hanging up in a tight squeeze, and was difficult to set up in the uneven terrain of a wild cave. I preferred to take my chances of finding a smooth ledge, top of a boulder, or a flat stalagmite on which to balance my camera. I seldom had any difficulty.

We had our trials and tribulations. We survived cave-ins, falling rocks, disorientations of both short and long duration, climbing disasters, and various encounters with assorted members of the animal and insect kingdoms. We knew the discomforts of bruises, scratches, and wounds both great and small. Bone-chilling water and perspiration-producing heat were a part of that way of life. Aching muscles and near-exhaustion came from overdoing a good thing. Often we wondered if what we found was worth the effort, and then some new-found underground wonder would burst upon us, and we realized that it was.

But most of all we had fun, which, I guess, is what it was all about.

To the best of my ability I have placed each feature and each room and each corridor in the proper location when describing the caves we explored. I sought aid in this from my old photograph albums, in which every picture is properly titled, described, and dated; in addition, I referred to a series of notes I compiled shortly after each cave trip. In spite of these precautions, I admit that there is a good chance of error. So, if I have placed one room where another should be, if I say I turned left down a corridor when in reality I turned to the right, if I describe a certain feature as being in one particular room when actually it is in another, I am sorry. Memory, at best, is a tricky thing—after forty years some details can become fuzzy.

I would be remiss if I did not thank at least two people without whose help I would never have finished this book. First, thanks go to my wife, Elizabeth, who endured. She endured the clatter of a typewriter into the wee hours of the morning; she endured my outbursts of irritation when things simply would not develop at the writing table; she endured my lack of concentration on things at hand while I mentally wrestled with an idea that would not emerge. And, above all else, she crawled out of bed many times late at night to bring a cup of hot coffee to ease my mental journeys into the dim realms of the past. Next, my heartfelt thanks go to my dear friend, Marge Nickelson, who helped iron out the rough spots in the manuscript and pointed out errors obvious to her but not to me because of my intimate involvement with the subject. And, most important, she gave praise and encouragement when I needed both badly. Without help like this, a writer is lost.

I took all the photographs, except those otherwise credited.

The photographs in this book are in the Photographic Scrapbook beginning on page 197. Each bears a number which will also be found in the margins of the text opposite that part of the narrative to which it pertains. When the reader sees such a number in the margin of the story, he may turn to the Photographic Scrapbook to find what is being described in the narrative illustrated in pictures.

1
Spider Cave

I SUPPOSE IT WAS INEVITABLE that Jim White, the first explorer of Carlsbad Caverns and a man who loved caves beyond comprehension, would be the one who directed me to my first wild cave. The thrill of traveling down that road on which he got me started will never leave me. I was sitting with him one afternoon on one of the benches in front of the entrance to Carlsbad Caverns in the spring of 1933. To the west and south the great loop of the Guadalupe Mountains shimmered in the haze, the aftermath of a dust storm the afternoon before. Jim's deeply lined wind-bronzed face and watery, fading grey eyes seemed as much a part of the desert landscape as the great limestone boulders around us. We had been talking about his early days in the great cavern beneath us, a subject always as near and dear to him as the morning sunrise.

"Jim," I said, "I know you've gone into lots of other caves in these hills. I surely would like to do some exploring in a cave that no one—except probably you—had ever been in before. Would you mind telling me how to find one?"

"Robert, if we had a strong wind from the south and you had a chaw of tobacco in your mouth you could almost spit into one right now."

"You don't mean it!"

"Yes sir. Right over the hill and down in little Garden Grove Canyon is a real nice little cave. I never went very far through it—too small to appeal to me, now. I got into several nice, small rooms. Some good curtain formations and a few stalactites. May be lots of cave there . . . I don't know."

"You wouldn't mind telling me how to find it?"

"Not at all. Just keep the country road at the foot of the hill where the highway starts up here. You can't miss it—there's a gate there. Follow the road for a mile or two until the first respectable canyon branches to the left. That's Garden Grove Canyon. Just walk up the floor of the canyon—you can't miss the cave, 'cause it's right in the bottom of the canyon."

A few months passed before we got a party organized to find the cave, but on July 16, 1933, five of us—Tommie Futch, Dave Wilson, Ted Fullerton, John Ruwwe, who was visiting with friends in Carlsbad, and I—left the car at the entrance of what we hoped was Garden Grove Canyon and started the hike up the boulder-strewn gulch looking for the cave. As Jim White had said, we couldn't miss it. We found the entrance, not much bigger than a man-hole in a city street, dropping straight down for about five feet, with a small hole leading off into blackness at the bottom.

"Sure hope we don't get a cloudburst while we're in there," Tommie commented, looking up the canyon. "Any damned fool can see where the water'll go if the hills start flooding." There was no arguing that point. If the canyon started flowing, this cave was going to take in a lot of water.

"What are you worrying about?" Dave snorted. "There isn't a cloud in the sky."

"Me worry?" Tommie retorted. "Just give me room."

He dropped down into the entrance and shone his light into the small hole leading off. Then he climbed back out and sat down on a rock.

"Nymeyer, this is your cave," he said. "You get first honors." I thought there was a guarded slyness in his tone.

I dropped into the entrance, hunkered down, and flashed my light up the tiny tunnel. At first glance I noticed only the smallness of the passage, and wondered if I could make it through. Then I saw something else. It was only a slight movement at first, up on the ceiling and down the upper half of the walls. Then my eyes accustomed themselves to the gloom, and I identified what it was.

"Not me!" I yelled, springing to my feet and leaping out of the hole. "You couldn't drag me through that."

What I had seen were spiders. Literally millions of them clung to the ceiling and walls, forming a black mat completely covering the rock. The beam of the flashlight seemed to irritate them, and they bounced up and down on their long legs until the entire tunnel seemed to pulse. Each tiny eye reflected the rays, and the entire quivering mass sparkled with millions of tiny pinpoints of light.

"What's the matter, Sam?" Tommie taunted, calling me by the nickname my father had given me many years ago. "They're only daddy longlegs. Everyone knows they don't bite."

"Everyone? Do the spiders know it?"

"Well," Tommie said, "you had your chance to be first. I'm going to have a look-see."

"I'm no piker," I replied. "If you go in, I'll be right behind you."

"Don't leave us out," one of the others said. "Who's afraid of a big bad spider?"

Tommie dropped into the hole and started inching his way out of sight. I followed right behind. This was a belly crawl for certain, and there was no way I could keep from knocking spiders off the ceiling. My hat scraped them off and they fell down my face, and I could feel them crawling in my collar. But Tommie was right—they didn't bite. He was grunting along laboriously, pulling himself forward with his arms, and by looking past him I could see the spiders banking up ahead of him in waves, trying to retreat from his light, until their very weight pulled them down from the ceiling in masses, which broke on the floor as the spiders scurried for the sides of the tunnel. I could hear the other men crunching along behind me, and knew from their curses that the spiders, rushing in panic-stricken flight, were pouring over them.

After about two dozen feet the passage gradually circled to the left, and the spiders were left behind as daylight disappeared. But it was still a belly-scratching crawl. At one spot the ceiling dipped so low that I became stuck, and it was only by rocking forward and back, digging up the loose gravel with my stomach, that I managed to go on. Loose, sharp-edged pebbles would bank up beneath me, adding to my discomfort as I dragged them along. Finally, after about fifty feet of this, the passage emptied suddenly into a room filled with delightful standing space. We were all panting and drenched with sweat.

We found ourselves in a chamber about fifteen feet square with a ceiling ten feet overhead. One large brown stalactite hung from the center of the ceiling, joining the floor to make a sturdy column that seemed to support the roof. A large number of smaller formations were grouped around it on the ceiling. A natural bridge, inverted like an upside-down rainbow, spanned the room from wall to wall. We noticed one corridor branching off to the right, and another led off to the left, beyond the bridge. We caught our breath and had a drink from our canteens.

"Well," Dave said in his Mississippi drawl, "let's see which way she goes."

We turned down the right corridor, walking erect, but were soon forced to stoop, and then to crawl. The passage sloped down, and the floor became a deep drift of sand that got wetter and wetter as we labored along. It showed ripple marks left by rapidly flowing water, and was littered with small pieces of driftwood, trash, and leaves. Sediment clung to the walls.

"See what I told you?" Tommie said. "I told you this place would run like a river if it rained up in the hills."

Finally we found ourselves stopped completely by mud and debris banked to the ceiling, so we turned back to the Auditorium, as we named the first room. Scrambling over the natural bridge, we started up the left passage. Stooping most of the time because of the low ceiling, after about forty feet we came to another chamber somewhat smaller than the Auditorium. A small pool of clear water nestled at the base of a dark brown mass of flowstone which formed a perfect frozen waterfall against the far wall, seeming to pour from a small tunnel near the ceiling. It was wet, and glistened under our lights.

"Water!" Ted hollered, and flopped on his belly to drink from the pool. We all followed his example, for the few sips from the two canteens we carried had only made us want more.

"That hole up there where the waterfall comes from sure looks interesting," Ted said, pointing it out with his light. "I think I can make it up there." He managed to clamber to the top of the flowstone, finding finger-holds in its fluted surface, and peered into the hole. "Nice cave back there," he called down. "Ceiling's only about three feet high, and the whole floor's covered with a pool of water. Stalactites all over the place. There's another room beyond the first one, with lots of formations and water there, too. I don't think we could make it into them."

We discovered a small tunnel leading down to the right, and crawled into it, squirming through a shallow pool of water near the opening. A few dozen yards of laborious crawling brought us to another small room in which there was just enough space for the five of us to stand. Two corridors branched off, and the right one led us, after several yards of stooping and crawling, to a series of four or five small chambers, all too low to stand in. One of these we named the Snow Room, because the floor was deep in pure white gypsum sand and its walls were winter white. Another larger room had several little alcoves in which hung white stalactites up to five feet long, covered with rough protuberances like popcorn. John accidentally struck one with his flashlight, and a bell-like tone reverberated through the small room.

"Listen to that!" he exclaimed. He tapped another, and a different note rang out. We gathered around and began tapping the stalactites, finding each had a different tone. Before long we were picking out crude resemblances to familiar tunes.

Dave completed a sloppy version of Dixie, then said, "We'll just have to call this the Music Room."

A deep, ragged fissure about twenty feet deep split one end of the Music Room. At first it seemed impossible, but I managed to find handholds on its rugged sides and made my way to the bottom. The others followed, and we found a small passage leading off, a continuation of the cleft. We crawled down it for a dozen or so yards, but it finally became too small to penetrate farther.

We went back to the upper level and returned to the room from which one corridor had led to the Snow Room. While the others rested, I crawled up the other passage, finding its floor covered with rich brown flowstone over which water trickled down the passage. I crawled up this tunnel, the seeping water saturating me to the skin. After about twenty feet I emerged into a small chamber with standing room. A corridor branched to the right, passed through two more rooms slightly larger than the last one, then ended in a blank wall.

I went back to the first of these three rooms and continued crawling up the wet passageway I had been following. Creeping around a sharp bend, I came to a sudden stop and felt the hairs rise on my neck. I had almost poked my nose into six sticks of dynamite, rotten and sodden, in a disintegrated paper bag. A coil of rotten fuse lay nearby. I had heard that dynamite becomes very unstable with moisture, and I wanted no part of this.

"What damned fool," I pondered, "would bring dynamite into a place like this? He could bring everything crashing down around his ears."

Fortunately, the tunnel was wide enough here that I could crawl around the dynamite, giving it a wide berth. Within ten feet the passage dipped to about eighteen inches of the floor. I poked my head and light beneath the arch and saw that the corridor led on, seeming to grow larger before it curved out of sight.

I decided against checking it out, as I had been parted from the others for many minutes, and was afraid they might be wondering about me. I headed back in their direction, and saw their lights flashing up the tunnel.

"Where in the hell have you been?" Ted groused. "We were coming after you."

"Cave exploring, what do you think?" I retorted. "And I believe I've found some more. Let's go see."

"No more for me today," one of the others replied. "We decided to find you and get out of here—it's been a hard day."

I was reluctant to leave because the expanding passage beyond the dynamite beckoned, but I realized they were right, and that we were still many crawl-weary yards from daylight. So we headed out, retrieving the string we had stretched behind us in places where we thought we might get confused on the return trip. As we crawled from the entrance and brushed the spiders off, the sun was low in the west, with long shadows stretching toward us from the rim of the canyon.

"There's only one name that fits this hole," I said, reaching down my collar and pulling out one of the critters. "Spider Cave it's gotta be."

The following weekend we were back. John Ruwwe had returned to his home in the east, but the friend he had been visiting, Jack Rasor, had joined us, and we picked up young Bill Bryan as a sixth member for today's exploration party.

"John really wanted to stay and come again today," Jack said, "but he just had to get back to his job. He really got a kick out of his cave trip last Sunday."

The crawl through Spider Gulch, as we named the entrance passage, was a repetition of the first trip. Bill took one look at the spiders and almost backed out. We decided not to go to the Snow Room area this time. I had impressed them with the possibilities of a bigger and better cave beyond the dynamite, so we headed immediately for that area. Tommie began to have difficulty with his lantern just as I started leading up the wet crawlway toward the dynamite.

"Sam, go ahead and see if you can get under the low spot you told us about," he said, "while I try to fix this damn lantern. We'll be right along."

I skirted the dynamite and succeeded in barely inching beneath the low arch beyond, ripping my shirt sadly in the process. Almost immediately I dropped off into a chamber with the ceiling well overhead and a dark corridor with plenty of walking room leading off to the right.

"Come on, fellows," I yelled back to the others. "I think we've got cave ahead."

"Be there in a few minutes, Sam. Hold up 'til we get there."

I sat down on a projecting shelf, shining my light down the beckoning corridor beyond. Faintly, far ahead, I caught the gleam of white stalagmites. And it was at that moment that something sneaked up from behind and bit me, something unseen and insidious, but potent, nonetheless. The bite was painless, and I wouldn't realize its import until hours, or days, or maybe even weeks, later. It was the cave bug. And it injected me with a cave-hunting virus that I was never to shake off. Through the rest of my years the lure of these dark, mysterious realms beneath the earth would remain with me, the thrill of anticipation of what might lie ahead around the next bend of the corridor, the utter quiet and peace and solitude, the unearthly beauties they held. To me there is more to crawling into caves than "just because they're there"; to me it is the promise of what they hold that lures me, the mystery of what might be found, the glistening wetness in the cool, quiet chambers, the fantastic charm of their decorations in millions of forms and types, the ever-present element of danger that might be encountered anywhere in the darkness. To me, there is no greater adventure.

The others soon joined me, squeezing under the narrow pinch-down and looking like a group of drowned rats from the crawl up the incline through the seeping water.

"This had better be good," one of them grumbled.

"This is gonna be good," another exclaimed. "Look at that tunnel go."

We started down the corridor, which was about six feet in diameter and swung gradually to the right. A sandy-colored, almost red, soil clung to the walls and carpeted the floor beneath us, but the ceiling was bare brown limestone. From above, scattered white stalactites hung in stark contrast to the red walls. Within a hundred feet we stepped into a small chamber some eight feet across, with a large array of white and tan formations decorating the ceiling. In the center of the floor stood three beautifully terraced, graceful brown stalagmites. The tallest one had joined with a snow-white stalactite from above. Each hanging formation bore a drop of water on its tip, and as we waved our lights about, the room sparkled like diamonds. The floor beneath our feet was soggy.

"This room made the crawl through the spiders worthwhile," Jack said softly.

"Yeah," Dave replied, "and I really think we're into some big cave, now."

But he was wrong, for within a short distance the corridor began pinching down, and soon we were on hands and knees, sogging through the soft mud. We were quickly covered with the sticky stuff, and it dropped into our hair when we brushed the roof of the tunnel. Occasionally we came to stretches where we could walk in a crouch, and even erect for short distances, but then had to crawl again. We covered probably more than a hundred yards in this fashion, the corridor twisting crazily about. We passed several smaller branching tunnels, but kept in what we judged to be the main corridor.

"Is this the 'big cave' you mentioned, Dave?" Ted asked, brushing some mud from his eyes.

Finally the corridor started growing larger and became a passage about twenty feet wide by half as high. The red soil still covered the walls and floor. We found places where it had crumbled from the walls, leaving the bare limestone showing. The deposit seemed to be about four inches thick on the average. We assumed it had been carried into the passages by flowing water, probably many years before.

Then we started encountering cave formations again, but how different they were from the normal type! They were mostly stalactites in low spots near the walls and usually connecting with the floor. All hung lusterless and dull, like white chalk. To the touch they even seemed softer than the regular formations we had passed. As a rule they were rather thick and heavy, forming in masses and clusters rather than individual stalactites. Against the red soil covering the walls they stood out in sharp contrast. Only a few small stalactites hung free on the ceiling. We came to one of the dead white clusters that, in profile, resembled the huge skull of a tiger, mouth open and fangs erect.

"Sure would like to have my picture taken in the tiger's mouth!" Dave exclaimed. So I got him to sit between the jaws, with Ted and Tommie alongside, opened the camera shutter on time exposure, and made about a thirty-second exposure by lantern light—we had not yet come to using flash powder. Up on one of the front tusks we noticed proof that we were not the first ones here, as someone had painted two letter A's.

"That must have been old Alcoholic Al," Bill muttered, "off on a binge and dragging his dynamite behind him." He had been very dubious about passing those sodden sticks on the way in.

As we walked along we found ourselves in a maze of branching corridors, most of them smaller than the one we were in. Then we came upon one of the day's strangest formations. A large cluster of dull white helictites grew out from the wall at a height of about four feet, twisting and turning in every conceivable direction; many, no larger around than a pencil, grew straight down for a foot or more, then seemed to go crazy and curl upward and about in loops like a

3

4

tortured worm. The mass reached to the floor. There were hundreds of the twisting helictites.

"Who was that gal in the old Greek story that had snakes for hair?" Dave asked.

"I think that was Medusa," Tommie replied.

"Then we oughta call these Medusa Heads—they sure look like snakes to me."

5

6

Many of the lower tips of the tendrils were buried in the red soil, sure proof that the dirt had come in after the formations grew to the floor. We found several more of these clusters as we made our way along. We came to a deep, narrow crevice cutting across the corridor. One side wall of the crack, from top to bottom, was covered with twisting white helictites, but the other side was red and barren. At one side the crevice was narrow enough for us to jump across, but we soon found our way winding along the brink of another ravine about fifteen feet deep and ten feet wide. Slender white stalactites hung from the bottoms of little protruding ledges and many helictites twisted about from the walls.

"That hole down there looks interesting," Jack said, indicating a small round opening at the base of the crevice wall.

"Let's go see," Tommie replied, always ready to try a new tunnel.

We managed to pick our way down to the bottom of the crack and stooped into the branching tunnel. It led into a small chamber with a steeply down-sloping floor.

"Black stalactites!" Dave exclaimed, pointing at the ceiling. Then we discovered they really were not black, but were simply covered with black mud. So were the floor and the walls. The sloping floor pitched steeply to a large, deep pool of water at one side of the room, but we could see the corridor extending on, at about eye level.

"Reckon we can get across?" Bill asked.

"I don't know," Dave replied, testing the mud. "It's slicker'n you know what."

Ted elected to try it first, and had gone about halfway across, skirting the wall where it joined the steep floor; suddenly his feet went out from under him and he started on a swift toboggan toward a sure bath.

"Yipe!" he yelled, grabbing out for anything to stop his catapult into the water, and fortunately caught onto a small stalagmite growing out of the mud. His feet were in the pool.

"For gosh sakes, Dave," he yelled, interrupting our hilarious laughter at his predicament, "throw me an end of that rope. You know I can't crawl up this slick bastard."

Dave pitched him an end of the lariat he had insisted on carrying into the cave, and we dragged Ted up through the mud. To call him a mess would have been a gross understatement. We gave up the idea of trying to gain the corridor beyond and crawled back out of the Mud Room, as we called it, covered with the black stuff.

Clambering back out of the crevice, we continued following the main corridor. Within a short distance we noticed a side passage, down which white gleamed in the beam of a flashlight. We turned into it, and came almost immediately into a room much larger than most we had seen. It was at least thirty feet across, with a ceiling twelve feet overhead. Down its center extended a row of white pillars, close together, and varying in diameter from the size of my arm to larger around than my body. A few tapering, sharp-pointed stalactites hung down among them, and short, round-topped stalagmites clustered about their bases. All were lusterless and dead; again I thought of chalk as I looked at them. We named this the Room of the White Pillars.

7

We continued down this corridor and were surprised to find that it circled back to the main corridor we had just quitted. Within a hundred feet we came upon more of the white columns, not as large as those in the room we had just left. They completely blocked the entrance of a small corridor branching off behind them. These columns showed character, for their bases bore a covering of sharp aragonite crystals, minute and clear and sparkling in our lights. They were a delicate, transparent, reddish brown, in contrast to the dull white of their parent columns. Dave leaned down and touched a mass of the crystals.

"What do you know," he muttered. "Just like grabbing a cactus." A few of the tiny sharp crystals had pulled free and were embedded in his skin. Several slender white stalactites hung down from the ceiling; they were no larger around than a baton, and sparkling crystals covered them.

We found a way around behind these pillars and crept into a

small room about six by eight feet, with a round dome ten feet overhead. It gleamed like a winter palace, with snow-white walls and ceiling, all covered with thousands of curling white helictites. Several slender, twisting, antler-like white stalagmites, like the bare boughs of a tree, reared up for three to four feet, adding to the eerie aspect of this tiny room.

"This would have to be the Fairy's Room," said Tommie, naming it.

Returning to the main corridor, we came within fifty feet to a narrowing-down of the passage as the floor started sloping sharply toward the ceiling, ending in a pinched opening about three feet wide and five feet high. Hanging down in the doorway were half a dozen slender white stalactites two feet long and as big around as my finger. Wet and glistening, each bore a drop of water on its tip. The longest wore a beautiful cluster of aragonite crystals at the end. The mass was as big as a football, and seemingly too large for the slender stem to support. The crystals grew in all directions, sharp and delicate, like enlarged frost etchings.

"This would have to be the Fairy's Wand," Tommie exclaimed, "if that was her room we just left."

Crawling through the small opening, taking great care not to touch the Fairy's Wand, we found the corridor immediately assumed its former size. And just beyond we came upon a large pool of water, cold and clear and refreshing.

"Best drink I've had all day," Ted exulted. "Sure feels good to get that mud out of my mouth. If it were bigger I think I'd just jump right in."

The edges of the pool were covered with thick deposits of the aragonite crystals, forming masses six inches high in places. I discovered their lethal character when I leaned down to drink from the pool and felt sharp pricks in my hand. I jerked it away and found it covered with the needle-sharp things, and each one I yanked free left a tiny drop of blood on my skin.

"Cactus Spring!" I exclaimed. "That's what we have to call this one."

The pool lay back under a low-hanging section of the roof. A large cluster of the Medusa head formations twisted down from the ceiling and penetrated into the center of the lake. Where they touched the water their tips ended in clusters of sharp crystals that appeared to be gradually growing up their length.

Around a bend of the corridor just beyond Cactus Spring we found an entire section of the wall for more than twenty feet covered from ceiling to floor with Medusa heads, the thousands of tiny calcite "worms" squirming in every direction. As we moved the beam of a flashlight rapidly across the face of the wall, the changing shadows seemed to throw the entire wall into frantic activity, creating a most bizarre and unreal effect.

"It's almost as though they actually moved," Bill muttered. "That's the damndest thing!"

Many corridors branched off in the Cactus Spring area, and the main corridor we had come down became just another of many inviting explorations. We walked and crawled several out as best we could, but soon became confused in their numbers, finding many that doubled back upon themselves. We were glad we carried plenty of string, as it always led us back to our spring.

"Fellows, I've had enough," Jack finally spurted, as we once again found ourselves at the end of our twine back at Cactus Spring. "Do you know how many hours we've been doing this? And it's a long crawl out."

We all felt inclined to agree, so we gobbled up the remainder of our lunch, washed it down with the cool refreshment from Cactus Spring, and started following our twine leaders out of the cave. We were careful to roll the string back onto our spools as we returned; we knew we would be back, and we did not want any guiding lines left that might confuse us on our next trip.

"Damned if this isn't just about the most miserable looking bunch I ever saw," Tommie exploded as we crawled out of the hole in the bottom of the canyon. And we were a sight. Clothes were torn and covered with mud, hair matted and grimy, skin scratched in many places and streaked with dirt.

"Boy," Dave exclaimed, taking a sniff at his underarm, "I sure would hate for that cute little blonde to see me now. I smell like a burned goat!"

"Well, anyway, it's one hell of a cave," Bill replied. "I wouldn't take anything for this trip."

It was to be nine months before we went into Spider Cave again.

Dave, Tommie, and I felt very professional about the whole thing. Just the week before we had hired out to take a couple of tourists from Toronto, Canada, Fred Burgess and Cal Livingston, on a trip through Endless Cave, and they had gotten such a thrill out of doing some "wild" caving that they chose to stay over an additional week in order to do some more. So we decided that today's wild cave would be Spider. The date was April 14, 1934.

We left town well before sunup, and the crests of the ridges were just turning golden as we walked up to the manhole entrance. I had wondered what the reaction of the two Canadians would be when they looked down the entrance crawlway and saw the spiders. It was about what would be expected . . . "Ugh." But they were game, and we were soon on our way, scraping gravel up with our bellies and spiders down with our hats. Once I thought Cal on the verge of bolting—to just where would have been a problem—when spiders filled his collar and several got stuck behind his glasses.

"That sure was enough of that," he burst out, almost leaping into the room at the end of the crawlway.

"What's the matter, Cal?" Dave chided. "I thought you liked cave exploring."

"Cave exploring I like," he answered in his soft Toronto accent. "Eating spidahs is some othah gent's cup o' tay."

We took them for only a brief tour of the section containing the Snow Room and the Music Room, then headed for our favorite area, which centered around beautiful Cactus Spring. We worked out a few of the intersecting passages as we went along, to be certain the tyros got a true feeling of exploring new places. The Canadians watched with interest as we strung out twine behind us whenever we started up a corridor about which we were uncertain.

"I notice you didn't leave string in many sections," Fred remarked. "You are most positive about the way ouwut of heah?"

"Oh, we've all been in here before," Tommie assured him. "No problem."

Our Canadians were most anxious to take back photographs to show their friends, perhaps so they could brag a little about their experiences in the wild west; so we explored in a leisurely manner, taking photographs carefully in one area and moving on to another while the smoke from the flash powder cleared. Our guests were particularly intrigued by the dead white columns in the Room of the White Pillars, and by the sharp aragonite crystals covering everything at Cactus Spring.

10

"Prickly little blightahs, aren't they?" Fred exclaimed on his first contact with them as he leaned over for a drink at the spring.

Suddenly Tommie looked at his watch. "Oh my gosh!" he exclaimed. "Look what time it is. And me supposed to take Mary to a party at four o'clock. Two hours. I'll never make it."

"Tell the lady we detained you," Cal said softly.

"She'll love that. Well, I'm on my way. You guys be careful."

"Don't forget, Tommie," I called after his retreating light, "we're taking Cal and Fred to Gunsight Cave tomorrow. You're going, aren't you?"

"Sure gonna try," he called back. "I'll check with you tonight after you get in."

"O.K. We'll start out of here in a couple or three hours."

Fred wanted to get a photograph of the Fairy's Wand if at all possible, so we wandered over there. Because of its location hanging high in the opening, it was not the easiest of subjects, but I finally got set up and fired a charge of powder. We found a particularly beautiful and delicate cluster of the white helictites we had come to call Medusa heads, so we stopped and set up a photograph there. In leisurely fashion we drifted back to Cactus Spring, stretched out, and ate half a sandwich and a candy bar.

8

"Well, fellows," Dave said, "it's almost five o'clock. We'd better be starting out of here."

"I'm afraid you're right," Fred answered. "It's been a delight. I want you to know this has been just about one of my best days."

"Well," I said, flashing my light at Dave, "lead on, Macduff. Take us out of here."

"Away we go," he shouted, and we set off behind him. He strode confidently along, stooped through a low passage or two, passed an intersecting corridor, hesitated a moment, then turned into it. In a few steps he came up against a blank wall.

"Whoa," he exclaimed. "Took the wrong turn back there. About face and follow me."

We went back to the mouth of this corridor. Dave looked both ways, seemed a little hesitant, then turned to the right, in a few yards strode past another branching corridor, then backed up and entered it. Soon we were crawling on hands and knees.

"Dave," I suggested, "I don't remember crawling this close to Cactus Spring."

"This sure as hell isn't right," he muttered. "Let's go back and take the other turn."

So we went back and took the other turn. In some fifty yards we came to a beautiful bank of Medusa heads along one wall. There was a hint of flash powder smoke hanging up near the ceiling.

"This looks most familiah," Cal offered.

"It should," I exclaimed, shining my light ahead. "There's Cactus Spring right up in front of us."

"Sam," Dave exclaimed, his tone a little anxious. "You reckon we got ourselves lost?"

I noticed the Canadians glance at each other, and saw Fred suppress a knowing little smile.

"Aw, we couldn't be lost, Dave," I answered. "We've been in here enough to find our way out. Let me give it a try."

I did not have any better luck at first. But finally, after several blind leads, I was in a corridor I knew was familiar. Striding with extreme confidence I forged ahead, stooped under a low section, and emerged in a narrow corridor that led to a four-foot climb over a slightly projecting ledge that shadowed the wall beneath. We crawled up into the corridor above the ledge and followed the passage for a few yards, where it made an abrupt turn into a blank wall.

"Well, I'll be damned," I exploded, not believing my eyes. I had been so positive I was on the way out.

"Let's go back to Cactus Spring and sit down and figure this thing out," Dave suggested. "We're just banging our heads against a wall like this."

After several branching corridors and one blind lead, I said, "So, where's Cactus Spring?"

But in a short while we found it, and sat down to drink from its cool waters and try to work out the maze we were in. Fred had lost his knowing little smile that said, "These guys are pulling our legs; they want us to think we're lost, to add to our adventure."

And so we would puzzle over the situation, work out a solution, start with great assurance on our way out, and come up against a blank wall. Or after trying several diverging tunnels, would come right back to where we started from. We made Cactus Spring our center of operations, working out from there. Neither Dave nor I had too much string, so we could not leave it marking all the corridors, but had to keep taking it up for marking new attempts. We scratched arrows in the floor, and before long arrows were everywhere.

Weary and worried, we gathered once again at Cactus Spring. We were all down to our last set of batteries. I looked at my watch . . . one o'clock.

"There's no use in us all going together, beating our brains out, and wearing out our batteries," I suggested. "Why don't two stay here and two try to work their way out? I think every one of us knows how to get back to here by now."

"O.K.," Dave said. "You and Fred stay here and get some rest, and Cal and I will give her another try."

They were gone about thirty minutes, and both Fred and I were sound asleep when they got back.

"No luck," Dave answered to our inquiring looks. "I wound up right back in that damned dead end where we crawl over the ledge. I'd have bet my mother's Bible I was on the right track."

So Fred and I set out. We came to a branching corridor I did not recall, and there were no arrows scratched on the floor leading out.

"Hey, Fred, this may be it. You stay here and I'll work it out. I won't go beyond the length of my string, so don't worry."

"Me, worry?" he joked, but there was not too much humor in his voice.

I started down the corridor, which made a sharp bend to the left after several yards. Soon I was stooping along under a gradually lowering ceiling. Then I was forced to hands and knees. I knew by then that this was not the way out, but "any old port in a storm" I thought. Just as the tunnel pinched down too tight for me to crawl, I noticed fine, lacy networks of roots hanging down from the ceiling.

And then I noticed the tiny footprints of animals, probably field mice, sprinkled liberally in the fine sand which now covered the floor.

"By God, I'm close to the outside," I exclaimed aloud. But my flashlight revealed ahead only a hole that nothing much larger than a mouse could have entered.

I backed out and rejoined Fred. He was sitting in the dark, saving his batteries. We went back to Cactus Spring. My watch said 2:30.

"There's a key to this thing," Dave exclaimed. "I know we've passed it a hundred times. It's right there, if we could just recognize it."

"What are our chances of getting out?" asked Cal, finally voicing the question.

"Oh, we'll get out. I expect Tommie's on his way here now. You know he was going to check with us about going to Gunsight Cave. Soon as he found out from Mother we weren't in, he'd be on his way."

"Let's all eat a bit of our lunch—better save a little," Dave suggested, "then take a thirty-minute rest. Then we'll try again. May as well turn off our lights, no use wasting them while we're resting."

We turned off our lights. Darkness and quiet settled in, broken only by our breathing and by the occasional dripping of falling water. Then I heard something.

"Listen," I whispered. Our lights flickered on, but we hardly breathed, straining to catch some sound.

"There," I said. "Hear it?" From far, far off a sound whispered faintly up a distant corridor. It was a human shout.

Fred jumped to his feet and yelled a response. I think we all yelled. Then we heard the shout again, louder this time, and soon a flicker of light splashed toward us. Then we could make out two figures coming our way down the very same corridor we had felt certain was the one that would lead us out. A loud curse rent the air.

"Where in the hell have you guys been?" It was Tommie, and his expletive-slinging companion was a friend, Bill Liddell.

"We've been here," I answered innocently. "Where all has you been?" My weak attempt at humor must have been contagious and must have offered relief from the strain of our last few hours, for everyone had a good laugh.

"I'll say this," Tommie put in. "You've got everyone plenty worried. When I called your mother to find out when we were starting for Gunsight, and she said you hadn't got in yet, I knew something was wrong. I imagined someone down here with a broken leg. What's been holding you up?"

"Believe it or not, but we've been lost."

"Lost?"

"We've been trying since six o'clock to find our way out of this cursed hole. And don't you laugh. It damned sure happened."

"Any goddamned fool that would crawl into a place like this ought to have his goddamned addle-pated head examined," Bill stormed. He was a noncaver, on his first caving trip, and was not hesitant about expressing his dislike for the whole idea. "Those sonofabitching spiders!"

"You all got anything left to eat?" Tommie asked. "I'm starving."

We split up what was left in our lunch bags, all took a good long drink from Cactus Spring, and Tommie set out in the lead to take us out. The way he took seemed familiar. Maybe a little too familiar. Soon he came to a ledge that shadowed the wall beneath, and beyond which, at a height of about four feet, the corridor led on. Tommie vaulted up over the ledge. Bill was right behind him, and I followed.

"Tommie . . ." I started to call. Then I turned a bend in the passage, and found Tommie and Bill gazing unbelievingly at a blank wall. Then the air exploded.

"Sonofabitch, I knew it," Bill yelled. "Lost in this spider-infested bastard. Sonafabitch, I told you so."

"Now, take it easy, William," Tommie soothed. "I can get us out of here. Just let me go back and get my bearings."

We retraced our way back toward Cactus Spring. I suggested that I go all the way to the Spring, and then have the rest move back down the corridor to the place where we first saw their lights coming in. That would certainly give us a definite point to work from. When their lights disappeared, I joined them. They were in the corridor which had just dead-ended. But this had to be it.

We walked up to the overhanging ledge, with the passage leading on four feet above. Just by chance I flashed my light, dim and failing, on the floor and along the wall beneath the ledge. It revealed a small hole, about three feet wide, leading down. Tommie saw it at the same time.

"That's it!" he exclaimed. "I remember now. We crawled up out of there."

All the time we had missed it in the shadows of the overhanging ledge and in our hurry to get on to the passage above.

The spiders were still in the crawlway out, and Bill never stopped cursing on the long drag to the surface. I really do not believe he repeated a word the entire way. The sky was silver with dawn breaking into the mouth of the canyon. A mockingbird singing up on the slope across the canyon never sounded sweeter, and the air never seemed fresher.

We got back into town just in time to get organized for the trip to Gunsight Cave we had promised the Canadians. And, surprisingly, they were raring to go.

"Bully fun," Fred exclaimed. "Lost, and all. Bully fun."

In September I enrolled at the University of Missouri, and while back home for the Christmas holidays I decided I just had to make a cave trip. So Tommie Futch, Dave Wilson, Glenn Hamblen, and I went to Spider to renew old memories. It was a great sensation just to feel a cave about me once again. We spent the day crawling about and prowling the old corridors.

The school year finally ended, and I was back in Carlsbad, working for the Cavern Supply Company at Carlsbad Caverns. A group of the young fellows who worked with me all lived at a dormitory on the hill, and I told them about the great little cave just "a tobacco spit" away. After that, nothing would do but that I take them to it, so one night, after work, we drove across the hill on a Park service road to a point overlooking Garden Grove Canyon. It was almost dark by the time we scrambled down the canyon slope and gathered at the entrance of Spider Cave.

Plenty of spiders still swarmed in the crawlway entrance, but there did not seem to be as many as I remembered from the previous year. Four of my working buddies were with me: Hershel Davis, C. W. "Sonny" Hagler, Roy "Sparky" Renfro, and Bacel Scott. I bellied into the passage, with Sparky right behind me and the others grunting along in the rear. I was not too preoccupied with the spiders, but I heard the others slapping and cursing behind me.

I had just made the bend where the passage swung to the left when I heard a soft whirrr ahead and caught the faintest whisper of movement before me and to the left. I froze and swung my light. A rattlesnake was just easing himself into his final coil not three feet away. His rattles rustled softly. He was not yet thoroughly aroused, just being cautious. I kept my light square in his eyes.

I was afraid to move; the others had banked up behind me and there was no way to retreat.

"What's holding up the party, Robert?" Sparky called.

I did not dare answer, for fear my voice might rouse the snake to action. I waved my foot up and down, stopped deliberately for a few seconds, then waved it again. I think my toe must have touched the ground and set up vibrations, for the rattling seemed to pick up in intensity. I kept my light right on the snake's head.

"Wait up, fellows," I heard Sparky call back softly. "I think Robert's in trouble. He may have a snake in front of him." I was amazed, but relieved, that he had so quickly interpreted my signal. I waved my foot up and down as though in assent to his speculation.

I have no idea how many seconds, or minutes, the snake and I were at a standoff there in that narrow passage. I never shifted my light from his eyes, and he stared back, tongue flicking in and out, his bells a soft buzz in the stillness. Finally his head wavered and his coils relaxed. Slowly he slid out along the ground to his right, and I kept his head always directly in the beam of my light. There was a small crevice at the base of the wall, and the snake seemed to flow softly into it and out of sight. His rattles disappeared last, still softly whirring.

I breathed a deep sigh of relief, and felt the sweat pouring down my cheeks. I realized my shirt was soaked.

"It was a rattler, all right," I called back," but he's gone now. Keep a watch in the crack to your left as you go by, but I don't think he'll bother you. He was as scared as I was."

Sparky called back the news and explained what had happened.

"You mean we're going on?" I heard someone ask, incredulously.

We went on. I took them to Cactus Spring and the Room of White Pillars, and showed them the fantastic Medusa heads, and

11

introduced them to cave exploration by stoopways and by creeping on hands and knees. They forgot the rattlesnake and the spiders in the unearthliness of this world we wandered.

On the way out I decided to detour by the crevice containing the Mud Room, which we had encountered on our first trip in. The big crack was still muddy, but not nearly as sloppy as we had found it before. We managed to follow it down for perhaps fifty yards before it became impassable. As we turned to retrace our steps, I happened to shine my light up and spotted a small tunnel opening near the ceiling thirty feet above. I could see white formations gleaming back behind the opening.

"I think I can make it up there," I said. "You all wait while I give it a try."

It proved to be a fairly easy climb, and the others soon joined me in the small room that opened behind the entranceway. Chalk-white columns, like those in the Room of the White Pillars, stood about, amid the same lusterless stalagmites and stalactites. The walls were white and dull. The room continued as a passage that wound about for fifty yards or so. In places it pinched down so narrowly that we could hardly squeeze between the walls, but plenty of head room remained above. Finally we emerged into a series of three chambers, the first two being about a dozen feet square and generously decorated with light-brown stalagmites standing amidst many small pools of water. Stalactites hung in profusion from the ceiling, and all glistened with water.

The last of the three chambers was the most startling. About thirty feet in diameter, with a ceiling ten feet overhead, it was almost completely filled with one huge chocolate-brown stalagmite twenty feet thick at the base and extending almost to the ceiling. The entire surface of this massive thing glistened with seeping water flowing down over its beautifully rippled face. Hundreds of winter-white stalactites, many no larger around than a pencil, hung from the ceiling, and many of these had grown down and joined the brown mass beneath.

"Just like a chocolate sundae," Sonny said.

"Or like Grandmother's pincushion stuck full of darning needles," Hershel put in.

12

13

We could hear water dripping, and found a small room beneath the great stalagmite. This room was almost filled with a pool of water four feet deep, which overflowed into a narrow crevice of unknown and unpenetrable depth. Whether the corridor extended beyond this room we never knew, because the great stalagmite completely blocked the way to further passage. Sparky, being the smallest, started squeezing his way between the stalagmite and the wall, but had to give up after about eight feet, emerging soaking wet from the water seeping off the Chocolate Sundae.

We made our way back to the Auditorium, and I told them about the Snow Room and the Music Room down the other corridor, but they decided they had had enough for one trip.

"We'll see those some other time," Scotty said. "Who knows, we may have to wait on his highness, the snake, and be a long time getting out of here."

But his highness was nowhere in sight as we inched our way out. He had probably long since forgotten the one-eyed monster with the brilliant orb that had faced him down in the tunnel, and now hunted mice along the rocky ledges outside—quite likely his destination when I had intercepted him.

The black sky was moonless when we emerged from the tunnel, and only starglow softened the darkness.

"How in hell are you going to find the car in this?" Scotty grumbled.

"I'm going to walk to it," I replied, confidence in my voice, but a lingering doubt in my mind. It *was* dark.

I couldn't believe it when I walked right up on the car, as though I had seen it all along.

"I don't believe it!" Scotty exclaimed. "No one could go that straight to something in this darkness."

"Someone with a perfect sense of dead reckoning can," I replied. I realized that the statement, if carefully analyzed, did not make sense, but it was about the only thing I could think to say.

I returned to Spider Cave on April 8, 1967, with a new generation of cavers and a regular guide from Carlsbad Caverns. The Park Service had erected a solid, circular wall around the entrance. It was about four feet high, seemed watertight, and had a locked manhole-type cover in the top. The wall seemed a good idea, as it would certainly divert flood waters out of the cave, unless it rained torrents. It might someday save some lives if a cloudburst deluged the hills while spelunkers were in the cave.

The spiders were gone. I suppose they did not care for the perpetual darkness brought about by the solid gating of the entrance; I remember we never found them in the crawlway beyond the point where daylight died. I missed them.

I found the cave about as I remembered it. The hard trip was made harder by the passage of the years. The dynamite was gone—I wondered what brave soul had dared touch it. Some additional signs of vandalism appeared in the form of red initials, dates, and some names on the white pillars. But evidences of broken formations and decorations were few. I did miss the Fairy's Wand which had hung in the distant opening between the passageways, and wondered who had dared try to get its delicate beauty through all the tortuous crawling to the entrance; I hoped whoever did it made it in good shape, since he felt he had to remove it—such fragile charm should never be destroyed.

One of my young companions showed me a tunnel that dropped in the area of Crystal Spring. They were then engaged in "pushing" it, and reported that it went and went, straight in the direction of Carlsbad Caverns. Who knows? Maybe someday the two caves will link up. But the new explorers reported mostly crawlways and stoopways, and such a discovery would serve little purpose, except for the thrill of knowing the two were one system.

It was good to renew my old memories of the cave where the caving bug first bit me—Spider Cave.

2
Carlsbad Caverns

A WRITER TODAY, contemplating a story on Carlsbad Caverns, has to face the task with trepidation. So much has been said about it, so many words have been written about it and photographs taken of it, and so many millions of persons from all over the world have seen it, one wonders what more can be done. But no story of caving in the Guadalupes would be complete without a section devoted to this great cave.

My first trip into the cave was made on August 20, 1926. I have always regretted that I did not make the trip while visitors were still going down via the old guano-mining hoist bucket, which operated several hundred yards east of the natural entrance. But by 1926 the wooden stairway, consisting of some 219 steps, had been built into the natural opening, eliminating the fearsome drop in the bucket and the tortuous walk through the unscenic bat-cave section.

I was overwhelmed by what I saw on that first trip into the great Caverns. Since that time I have made at least fifty complete tours, and during two summers in the mid-thirties I worked for the concessionaire at the cavern, going daily into the cave by elevator to the lunch room. I have never failed to be overwhelmed each time I have gone down. There is just no describing the fascination of the place. One author summed it up nicely by calling the cave simply "Incredible Carlsbad." During the two summers I worked at the caverns, I think the two most common expressions I heard from the tourists were "I can't believe what I saw," and "It all seems like a dream."

More than all the other stories written about the early history of the caverns—the Abijah Long story of *The Big Cave*, the Jim White story *One Man's Dream*, the *National Geographic* story by Willis T. Lee, the Earnest Frank Nicholson story, partly fact, mostly fantasy—I think the fragmented stories of Ray V. Davis, the least publicized of all, deserve mention here. No one can deny the significance of his early photographs to the recognition and development of the caverns. A farm boy from Kansas, he became involved in photography with the purchase of a two-dollar box camera. This led to the establishment of a "Photograph Gallery" in Carlsbad, which helped fill in the hard times when the crops failed.

Jim White, the cave's first explorer, first took Ray into the caverns, about 1915, with the hope that some photographs could be made to prove Jim's stories about the magnificence underground. Many people had scoffed at his tales, remarking that he was as "batty as his bat cave." It was on this first trip with Jim that Ray's determination to photograph the cave began.

15, 16   He had had no experience of any kind in cave photography, and
17, 18   his was strictly a story of terrific labor and trial and error. That his
19, 20   early photographs were as good as they were is in itself fantastic. From simple hand-held flares of the type used on the railroads, to flash-powder guns, to an alcohol-magnesium burning floodlight which he designed and built himself, he progressed from single illuminated scenes to those lighted by many bursts of light in widely separate and carefully selected sites. These multi-flash shots gave depth and illumination to his cavern photographs that had never been attained before, and are seldom equaled today.

In those early days the trip into the cave was a rugged venture over primitive trails and precarious handholds. A small sack lunch and a lantern were burden enough. Consider, then, the stamina of this man, whose favorite camera was an 8 × 10 inch job that weighed about sixteen pounds. A home-made wooden tripod accompanied the camera and outweighed it by thirty pounds—it had to be sturdy for the duties that faced it. He usually carried at least two other smaller cameras, a 5 × 7 Graflex that was no light number, and a folding, twin-lens job for taking stereoscopic pictures. Add to this a duffle bag holding his magnesium-alcohol flare, supplies for it, cut-film holders, tripods for the smaller camera, and various other items of equipment he might need. Even with the help of a laborer or two to aid in transporting the equipment, he faced a formidable problem just getting to the site to be photographed. It is a fitting tribute to Ray that, even now, notwithstanding today's flash bulb and strobe flash techniques, high-speed films, and the enchantment of color, his magnificent black and white photographs are seen and admired in publications and collections throughout the land.

On that first trip of mine into the cave in 1926, I carried a little folding vest-pocket Eastman camera. It had an adjustment for time exposure, but it was a newly acquired toy, and I was happy just with snapshots in the sunlight. I had to content myself with a picture looking out the great natural entrance. On my next trip down, June 2, 1927, I did better. I had mastered the technique of the time exposure, and managed to take a few respectable timed shots in the electrically lit portions of the cave, notably in the King's Palace.

By the time the next year rolled around, I had gone to work for Ray V. Davis, doing odd jobs around his studio and building up a determination to become a photographer. Ray decided he needed some new shots in the caverns, and on January 15, 1928, he took his young son Eugene, his brother Kenneth, and me on a photographic expedition down below. When I saw the stack of equipment he proposed taking, I knew why he had asked the three of us to come along. And it was on this trip that I was initiated into the intricacies of cave photography, and learned the advantage of multi-flash exposures over photos taken with a single source of illumination.

I also got to see the Rube Goldberg contraption Ray had designed and built for lighting his scenes. It seems a miracle now that at some time in his career he did not blow his head off. His "gun" was designed around a trough, or pan, similar to the old flash-powder pans that news photographers used for indoor shots, only the bottom of Ray's pan had a thick layer of sponge. A metal tube about the size of a pencil was stuck up through the bottom of the pan near the center and welded in place. It protruded up through the sponge. Attached to this tube where it stuck out beneath was a length of rubber hose which extended down and entered a metal container of about a pint capacity. The hole in the top of the screw lid of this container was small enough to make a tight seal around the rubber tubing. Another tight-fitting

21

22

hole in the lid of the container held another length of rubber tubing long enough to reach Ray's mouth. A foot-long length of conventional water pipe, welded to the bottom of the pan, served as a handle.

It was with a great deal of curiosity that I watched Ray get ready for his first photograph. He poured the pint container about three-fourths full of magnesium powder, set up two or three cameras, cautioned us about showing lights in the camera field, opened the shutters on time exposure, and prepared for his first flare. From a can he carried in his pack he soaked the sponge with alcohol and applied a match. Then, holding the contraption high overhead, he blew into the mouth tube, forcing the volatile magnesium powder up through the other tube into the alcohol flame, and a brilliant, blue-white flare blazed up. The darkness vanished like magic. He could control the amount of light needed by the length of time he blew the powder onto the flame. He walked about, carefully lighting his way with a shielded flashlight, and at pre-selected spots where rocks or stalagmites hid him from the cameras' view he would blow another flare. Then he wended his way back and closed the shutters.

"That's quite a light you've got there," I remarked.

"Best thing for cave photography I ever saw," he replied, "even if I did make it myself."

"What would happen if you forgot and inhaled sometime, sucking that flame down into the can of powder?"

"I probably wouldn't be around for the next picture," he answered dryly.

This was the first of several photographic expeditions I made into Carlsbad Caverns with Ray V. Davis. One of the trips which comes vividly to mind was made in order to photograph a special party of V.I.P.'s in the King's Palace. Since our plans included taking only the one shot, we were not burdened down with equipment—Ray carried the 8 × 10 camera and I shouldered the heavy tripod. We took the photograph without incident and started the long wearisome trudge up the inclined trails, with here and there some flights of wooden stairs, to the entrance a mile and a half away and 800 feet above. We had covered only a fraction of the distance when, without warning, the electric lights went out, and we came to a sudden halt in total darkness. The electrical system, at that time, was powered by a generator on the surface, and it was not immune to failure.

"You got a light?" Ray asked.

"Gosh, no," I replied. We had figured that, with the electric illumination, there was no use burdening ourselves with unnecessary lights.

"Oh, well, the lights will probably come back on in a few minutes."

"They'd better not take too long," Ray grumbled. "I have an appointment for some wedding pictures this afternoon." We waited and waited in the darkness that has no peer.

"I've got to get out of here," Ray said finally. "Feel around in the pack and see how many matches are in the box." I found the container and was happy to report it held a brand-new box.

"O.K., we're going to get out of here by match light."

"That will be something to talk about," I commented.

"Here's what we'll do. You take half the matches. I'll strike one, and you run as far up the trail as you can see, or until the match burns out. Then you strike a match, and I'll run to you, and farther if the match holds out. I think we can get to daylight before we run out of matches."

I had my doubts, but was astounded to discover how much light a single match could make in such complete darkness. I am certain that on many runs where the way was fairly straight and even we covered a hundred yards in one sprint while a single match burned. But the matches were getting pretty low when we caught the first glimmer of daylight filtering down through the great entrance. I could not remember it ever looking so good.

I remember another picture-making trip when a state convention —I believe it was the Masons—was held in Carlsbad, and some ceremonies were scheduled for down in the caverns. They contracted for Ray to make a group picture down inside. He decided the group was too large to photograph successfully in the King's Palace, so they moved out into the Big Room area, pretty close to where the underground lunchroom is now located. When shooting pictures containing people, Ray resorted to conventional flash powder, as it produced a fast exposure and eliminated any subject movement that might occur during the slower-burning magnesium flare process. He had a flash pan arrangement for this, triggered with a spark—a cigarette lighter on a giant scale. I was to fire this after he got the big camera set up and focused.

Not satisfied with what he saw on his first look through the ground glass, he picked up the heavy camera and tripod, and forgetting that he was not on the familiar smooth floor of the King's Palace where he usually shot such pictures, started backing up. I glanced around to see how he was getting along just in time to see Ray, camera, and all disappear like a rabbit into a magician's hat. He had backed off into a hole somewhat deeper than he was tall. The clatter of his tumble echoed about the great chamber, and an excited ripple of voices came from the waiting group.

I ran over to see if he was still alive, and reached the hole just as his head came over the rim.

"Are you all right?" I asked anxiously.

"I think so," he answered, "but I'm not sure about the camera."

Examination revealed that he had survived the fall with only a few scratches here and there, and the camera was sound and serviceable. The big tripod had not fared as well. One wooden leg had snapped off, and the thumb screw into the camera socket was badly bent. It took several minutes to calm the crowd down and get them arranged again. By putting the big camera case under the broken end of the tripod we were able to get a fairly level stance for the camera, and managed to shoot the picture. Out of sympathy for Ray every member of the convention bought a copy of the photograph. He decided the trip had been worthwhile.

In the spring of 1929 a hot-shot movie promoter hit town driving a big black Cadillac emblazoned with the words "Jack Irvin Productions, Hollywood." With a beautiful wife and a small retinue of yes-men, he registered at the best hotel, tooted a horn of glamour for the certain fortunes to come to all believers, received plenty of free publicity in the local newspaper, and proceeded to sell stock in the movie he proposed to produce, all to be filmed locally. He found plenty of believers.

With the necessary capital raised and banked, production got under way on *The Medicine Man*. All filming was to be done in the Carlsbad vicinity and in Roswell, where Irwin also found plenty of believers. The climax of the movie was to be filmed in Carlsbad Caverns. Irvin did not do too badly on the cast, getting old-time favorite Tom Santchi for the lead, with portly Jean Layman playing his faithful servant in blackface, and city-slicker Philo McCullough sneering his way through the role of dastardly villain.

The scenes in Carlsbad Caverns were filmed at night after the regular daily tourist visits were over. Although they were to be the last scenes in the movie, in typical Hollywood style they were the first ones filmed during production. Ray V. Davis had been retained as technical advisor for the underground shooting, and on the night of June 28, 1929, he got me a pass as an assistant to view the final night of shooting. In spite of the typical slow methods on a Hollywood set, with adjusting of lights, instructing of actors, yes-manning of director, and retaking, retaking, retaking, the scenes were finally shot. Despite all the folderol, I found the evening exciting. When the final "Quiet!" had been shouted, and a hush had settled in the great chambers, when the torches were lighted and the actors got into their lines, the magic of Hollywood took over and it was easy to believe the story being told.

The finished movie actually was not too bad—the story was believable, and the acting was good. Unfortunately, it was completed as one of the last silent films when "talkies" were taking over. The production had run into difficulties and was finished far behind schedule; Irvin had run out of money and gone deeply in debt, and there were no finances available to dub in a sound track. The finished product hardly made an echo as *The Jazz Singer* brought Hollywood to its feet. The believers lost their shirts.

In the summers of 1935 and 1936 I worked at the Caverns for the concessionaire and lived on the site in a makeshift boys' dormitory located out of sight of the tourists on a hill north of the natural entrance. After the last visitor had left the cave in the late afternoon and the evening meal was behind us, it was not unusual for a group of the boys to wander down to the great opening and wait for the evening bat flight. At that time the bat population using the "bat cave" section was estimated in excess of three million, and the sight and sound of their emergence from their daily deep sleep underground were nothing short of magnificent.

Having seen the event many times, we were interested not so much in the bats as in the tourist girls who gathered to witness the spectacle. Many were eastern girls who felt a deep curiosity about "westerners," and quite a few were eager to become better acquainted. Only a twenty minute's drive to the east was Black River Village, a large tourist court complex, with wonderful swimming facilities in the river, rolling grass lawns for picnics, and a place for dancing to a juke box. It was easy for a romance to start budding in such surroundings, and several of these eventually culminated in marriage—all because bats flew from the great caverns.

POSTSCRIPT

Carlsbad Caverns National Park still ranks as one of the prime scenic attractions in the world. Over twenty million visitors have trod its great halls and gaped in amazement at the beauties its vast auditoriums hold. In the largest of these, the Big Room, the cave loses its identity as such, and seems more like a great canyon with a roof. In many places the ceiling is almost 300 feet overhead. It is shaped like a huge cross, with the stem over 2000 feet long and the crossbar stretching almost 1200 feet. This is the room of the giant stalagmites. The Giant Dome towers over 60 feet, with the great twins standing nearby, like Apollo and Artemis, but called simply the Twin Domes. White and glistening the Crystal Spring Dome still grows beneath a mantle of water dripping from above. The Temple of the Sun, perhaps the most symmetrically beautiful formation, rises to a height of 42 feet. Best known of all, the Rock of Ages stands atop its own little mountain and leans against the cavern wall, as though for a better view of the great chamber spreading out in every direction before it.

The King's Palace, the Queen's Chamber, and the Papoose Room, less staggering in size but more intense in intimate beauty, leave the visitors speechless, or whispering in awed wonder, as though fearful of disturbing the deep feeling of utter peace that pervades the air about them.

The walk into the cave has set the stage for all of this. A leisurely stroll down the inclined trail beneath the sweeping arch of the great entrance is followed by a descent of about 150 feet to the main cavern floor. Here the unscenic Bat Cave leads off almost straight ahead, while the main trail makes an abrupt turn to the right, dropping gradually for another 200 feet into what Jim White called the Devil's Den. From here the way pitches steeply, but still by smooth trails, for some 375 feet to reach the comparatively level floor leading to the Green Lake room and into the chambers of royalty.

Within the last year the Park Service has drastically changed its method of showing the Caverns to the public. The tours are now self-guided. The visitor enters at any time he chooses during the daylight hours when the cave is open. He strolls along at whatever pace pleases him, and stops as frequently as he desires to view the wonders or simply to rest. He may take as many photographs as his budget will allow, flash or time exposure, provided he does not interfere with others' enjoyment of the trip. And now he can have the sights explained to him by means of a light-weight receiver and headphones issued when he buys his ticket; as he leaves one broadcasting zone and enters another, prerecorded descriptions and information are transmitted by strategically spaced transmitters and picked up as he walks along. Park Service Rangers are stationed along the trails to protect the beauties from harm and to offer additional information.

Visitors taking the regular Park Service tour of some three and a half miles seldom know that the part they see comprises only about one-tenth of the Caverns' total surveyed passages. Branching corridors along the trail lead to more and more cave, dark and undeveloped. Some plunge to greater depths, often surpassing 1100 feet, and others climb above the traveled portion, often for several hundred feet, to hidden rooms of rare and delicate beauty.

In the past two or three years, I have made trips to some of these primitive areas as a guest of the Park Service. One of these was a night trip to a great new section, discovered in June of 1966. Called the Guadalupe Room, it is excelled in size and magnificence only by the

Big Room known to the tourist. The trip to this section is harrowing in the extreme, with much of the distance covered by crawling or stooping. One particular spot, called Matlock's Pinch after one of the original discoverers of this section, is no place for a claustrophobic.

"I don't know, Mr. Nymeyer," one of the young members of our party said as we prepared to negotiate the pass. "It's going to be pretty tight for you."

"Implying that I have grown fat and soft," I retorted.

The snug squeeze drops down, under, and up in a tight U, so close that I found I could not make the turn by starting on my belly, as my vertebrae would not arch backwards enough to make the bend; I had to lie on my back and drop down head first, arms extended, inch my way beneath the foot of the U, and then find fingerholds on the opposite side to enable me to pull myself under and up to the crawlway beyond.

"The young man was right," I thought in near panic when halfway through the ordeal. "You have grown fat and soft, and an old fool like you has no business in a place like this."

But finally to gain the walking corridors, pass the magnificent White Giant, like ice and growing, to stand on the high balcony and look out into the vastness of the great chamber made the arduous trip worthwhile. Our flashlights were like fireflies in a garden at night, promising much and giving little. The tortuous drop down the steep incline to the floor of the gigantic room is treacherous, as evidenced by an occasional clatter and a grunt as someone's feet left the terrain. But the beauties so rampant below are payment enough. Water drips everywhere, stalactites grow by the millions, cascading waterfalls of stone decorate the walls, and small, graceful stalagmites stand about like guardians. Brilliant colors, particularly lemon yellows and tangerine oranges, blaze throughout. In one small room adorned with varicolored stalactites in uncountable multitudes, a perfect yellow canary of stone perches in solitary splendor, head upraised, as though ready to burst into song.

"So much beauty," I thought. "So much virgin, unspoiled beauty. What a pity that the public may never see it."

On another night excursion we made our way down the Left Hand Tunnel, which branches off from the underground lunchroom, down narrow but high-ceilinged passages, past stately stalagmites guarding the trail and dark intersecting corridors that invite exploration. Our way led steadily down, to end finally at a precipitous drop which our flashlights barely penetrated. A heavy hawser was securely tied at the top and dropped over the lip.

"How far down is that?" I inquired, thinking of muscles already weary.

"Only about two hundred feet," an eager young explorer flipped off lightly.

The drop was not a free fall, but it was just about the next thing to it. I did thank my stars for occasional ledges offering rest stops. At the halfway mark a wide shelf gave me momentary relief and sitting room to recover my strength. Far down at the bottom, water gleamed a beautiful sea green in the light of a gasoline lantern placed there by an explorer who had preceded me. Then I heard a gigantic plop, and echoes roared up from the depths. Another followed, and I heard a shout, and was certain disaster had struck the leaders. Then a gale of laughter and shouting exploded upward.

"What is going on?" I asked a companion resting alongside.

"They're taking a swim in the Lake of the Clouds," he answered.

Swimming is one of the things I enjoy most, and with perspiration running in rivulets and clothing sopping, I knew this had to be for me. I managed the last hundred feet down the hawser without disaster, staggered to the rim of the beautiful pool, gaped once in awe at the great mass of roiling white flowstone dropping down and hovering over the water, and fell into the lake, clothes and all.

A dip like that will revive your vigor, or at least your spirits. After the first shock of the cold, my swim in the underground lake was one of the most exhilarating I have ever taken. Overhead, ominous and threatening, like a great, boiling thunderhead, the "cloud" from which this room received its name dominated the scene.

"How far down are we?" I asked the park ranger with us.

"We are down eleven hundred and thirteen feet," he replied. "This is the deepest part of the Caverns, as far as now known."

"How deep is the lake?"

"Right over there at the deepest part it is twelve feet."

I swam over to the spot indicated, the water a dark green beneath me. Then I upended, swam down, and firmly pressed my hands on the bottom. I had touched the deepest known part of Carlsbad Caverns.

We prepared to leave this most enchanting of rooms. I stopped at the bottom of the rope, and saw the younger explorers scrambling up with youthful abandon. A feeling of despair crept over me. I had serious doubts that I could make it. At the halfway ledge I stretched out, panting, with the sweat again streaming down and weariness in every bone. Since that trip I have read several discourses on hypothermia and the danger it presents, especially to cave explorers and high mountain hikers. Had I known of it then, I would most certainly have succumbed instantly on the spot to its perils. As it was I let myself have it with some pretty strong thoughts. "You damned old idiot! You ought to have better sense than to get into a jam like this. You think you're still a kid? You gonna let these youngsters show you up? You got yourself into this mess, so get up on your feet and move up that rope."

And I did.

I had felt the same way the night we went to the Guadalupe Room. I remember vividly the complete incongruity of the scene, as I bellied through the last horizontal crevice, boneweary and desperate, and saw, twenty feet below me, the smooth paved trail used by the tourists on their daily guided trips through the Caverns. I wondered if I could possibly make that last twenty feet. Later, as I walked, numb, along the pavement, to emerge finally into the glare of electric lights and stagger into the sterile, metallic square of the elevator car and rise in one minute to the surface 750 feet above, I felt it was all unreal. The trip into the wild, untamed splendors of the virgin cave just beyond the blaze of electric lights seemed only a dream, a jigsaw puzzle whose pieces would never all fit together into one completed picture.

They never have.

3
Black Cave

I HAD FIRST HEARD OF BLACK CAVE from Carl Livingston, pioneer resident, writer, and historian of the Carlsbad area, and one of the first to explore the Guadalupes solely for their beauty.

"Why do you call it Black Cave?" I had asked, for to me all caves were black until someone shone a light on them.

"Because it's a black cave," he had replied. "Someday we'll have to go there and I'll show it to you. Then you'll know." But Carl died before he found the time to show me the cave.

Dave Wilson, Ted Fullerton, and I had spent two separate weekends crawling and walking the mazes of Spider Cave, where we had been bitten by the cave-exploring bug; so on a hot August 6, 1933, we headed out before daylight for the Shattuck ranch atop the Guadalupes, where we picked up Julian Shattuck, who knew the location of Black Cave.

We drove the flivver as far as Dark Canyon Fire Lookout, where the road ended, and set out afoot for the cave, which Julian said was about three miles to the northeast. It was a hot three miles, and seemed like ten. It was one of those rare days in the mountains when the sun bore down from a cloudless blue sky and not a breeze rippled the pines. We were dripping sweat long before we struggled up the final slope from the canyon floor to the cave's mouth, a small hole hidden at the base of a steep wall of rock.

"It's bound to be cooler in there than out here," Ted panted. "Let's get inside!"

We had to stoop to get in the opening, but the straight corridor soon enlarged sufficiently for us to proceed upright. The cool air revived our lagging spirits, but our lights, after the bright sunlight, revealed very little. Proceeding carefully over the uneven floor for some sixty yards, we ducked under a low arch and stepped out into a long auditorium which seemed to be about a hundred feet wide. Our lights revealed the ceiling arching about thirty feet overhead, but details were dim.

"Damn it!" Dave exclaimed. "These were supposed to be brand-new batteries—I just bought 'em yesterday."

"Me, too," I answered, shining the weak beam from my flashlight about. It picked up a large stalagmite coming into view as we walked along. As I moved the ray about, other monoliths came into view, but disappeared into blackness as soon as the direct spot of light moved off.

"Damn it, fellows!" Julian exclaimed. "Our batteries aren't weak. You know what's wrong? Everything in here is black as Hell; it's like trying to spotlight a black cat in a coal bin at midnight."

And he was right. The stalagmites, the walls, the ceiling—all were black. They caught the beams from our lights and absorbed them. No reflections came back to illuminate the scene. All the cave—even the floor—was a dull, dead black, broken only by a glitter where a falling drop of water cast a reflection from an ebony surface. Closer inspection revealed that the black shaded into a dull heavy brown or a dark grey in many places, but where were the gleaming whites and tans we had seen in Spider Cave?

We picked our way carefully into the auditorium, and soon realized we were walking along a narrow ridge, or hogback, that split the room squarely down the middle. Narrow chasms separated the hogback from each wall. The blackness of the place made judging distance difficult, and it was likewise hard to estimate the depth of the canyons on each side. In many places we could see that they were only eight to ten feet deep, but in others they would drop to uncertain depths that could have been as much as fifty feet.

"I feel like I'm on the edge of the Pit," Dave grunted as he eased around a crack in the hogback, which was dotted with pits of varying depth. The going became quite precarious—holes looked like shadows, and shadows looked like holes. A few small stalagmites stood here and there on the ridge, somber gnomes guarding the passage. Occasionally the lights picked up a gleam from a stalagmite covered with dripping water, the moisture throwing back reflections.

Suddenly there was a flurry of sound from behind me, muffled curses, a clatter of falling, and the metallic ring of something clanging down into the chasm to my right. Dave and I whirled and our lights revealed only Ted's head and shoulders above the edge of the chasm, one arm hooked about a small knob perched on the rim. His face stood out stark and white in the darkness! Just as Julian, who was bringing up the rear, reached over to give a hand, Ted let go and disappeared over the edge. But there was no sound of falling.

"Sonofabitch!" came up out of the darkness. Then a short laugh. "Jeez, I thought I was a goner. Then my feet touched the floor. It's not deep here at all." He retrieved his flashlight and scrambled up to join us.

We continued on down the hogback, and soon it merged with the side chasms, and we were walking along a level floor. Then our lights began picking up occasional splotches of color, breaking the monotony of black. Here and there a stalagmite wore a crown of deep yellow or orange. Deep rusty browns stood out. The cave was getting wetter, and nearly every formation glistened in our lights. We could hear water dripping all about.

Then we came upon a huge stalagmite that curved up in graceful lines, forming a dome that almost touched the ceiling. In fact, the stalactite from which the water had dripped to form the dome had grown down to meet it, merging to form a column eight inches in diameter connecting the dome with the ceiling. The big dome was of the same dark, lifeless color—a mixture of black and brown. A smaller one about six feet high nestled at its feet, and another large one only a score of steps away—almost a counterpart of the big one—reared massively toward the ceiling.

"Ought to call 'em Papa Bear, Mama Bear, and Baby Bear," Dave grunted. But we settled by just calling the largest one the Big Dome.

The farther we went into the cave, the more water we encountered, and as the dripping water increased the black stalagmites decreased in number, giving way to ones of brighter hues. Lighter browns, mixed with oranges and lemon-yellows, began to appear.

"Hey! Here's a lead," Julian exclaimed from over to one side, against the wall. A small hole dropped steeply down at the floor's edge. We eased our way down and through, dropping into a small room about thirty feet in diameter. It made a strong contrast with the main cave above, seeming almost a riot of color. Many stalactites hung from the ceiling, tinted with bright yellow and sienna and orange and brown. Similarly colored small stalagmites squatted about on the floor. Nearly all the formations glistened with falling water, and little pools gleamed on the floor.

"Persephone's Palace," I murmured.

"What are you muttering about?" Dave asked, wiping his mouth after a long drink from one of the pools.

"Just remembering my Greek mythology," I answered. "Pluto, the god of the underworld, stole the beautiful Persephone and imprisoned her in Hades, building her a palace in the depths of darkness. Don't you see? Here is her palace, light and glittering and gorgeous, while in the cave above is the blackness of Pluto's hell!"

Dave looked at me a moment. "Sam," he said slowly, "the place has got to you. We'd better get you out of here."

We crawled back up into the main part of the cave and continued on, but soon found it growing smaller. We estimated the entrance was half a mile behind when we reached the end of the cave. We searched carefully for any hidden tunnels leading off from the auditorium as we picked our way back toward the entrance, but we found none.

Near the entrance, in full view of the blue daylight streaming in, we found a smooth stretch of wall where many persons had smoked or scratched their names. Many had included the date of their visit—I remember some dated in the late 1800's.

"Look!" Julian exclaimed, pointing. "There's Dad's name." There were many we recognized from the area's early history—Shattuck, McCollum, Thayer, Middleton, Lee—and many we had never heard of. Not to be outdone by the old timers, we left our names on the register before we left the cave.

The blazing sun had cooled scarcely at all, and its brilliant white light blinded us for an unduly long time, after the somberness within.

As we stepped out, Dave paused and looked back into the darkness. "Blacker'n the shadows of hell," he said half to himself, shaking his head.

4
Hell Below

"WHAT THE HELL AM I DOING DOWN HERE!" was one of the first thoughts that crossed my mind as the echoes of the crash of my descent rumbled about in the vast darkness and died away to a whisper far overhead. Then there was only a great silence and the occasional sound of dripping water. And the darkness.

It had all begun some months before—on August 6, 1933, to be exact—when Dave Wilson, Julian Shattuck, Ted Fullerton, and I were headed back across the ridges for our car parked at Dark Canyon Lookout, following a visit to Black Cave. It was a typical dog-day in the Guadalupes, the sun hot on our backs, the distant scarps soft and hazy with a promise of Indian summer.

We stopped for a rest and a swig from the canteens on a high crest. Suddenly, almost from nowhere it seemed, a short Mexican in faded work tans and with a red bandana binding his hair walked into our midst. A small black and white dog trotted at his heels, and dropped panting in the shade of a boulder.

"Hi," the stranger breathed softly in greeting. "She's plenty hot today, no?"

"Plenty hot," we agreed.

"Wat you do up here this time year? You look for deer to hunt soon, maybe?"

"We're not hunting deer," Julian explained. "We've been looking for caves."

"Ah, cuevas! You like caves? They pretty dark!"

"We like to explore them," I explained. "What are you doing up here?"

"Me? Me and Perro, we look for damned lost goat. You see stray goat, maybe?"

I had a faint recollection of having heard a goat bleating on the ridge above Black Cave, and told him so, pointing out the ridge in the distance.

"Stupid bastard," he grunted as he lurched to his feet. "Maybe so I let him stay lost. El Patron got plenty damned goats, nohow."

In a moment of inspiration, I asked, "You haven't run onto any caves in your goat-herding over these mountains, have you?"

"Ah, si," he replied happily, glad of an excuse to delay the search for the lost goat. "Just last week, I theenk . . . over yonder." He pointed vaguely to the south. "A hole in the hill. Not very beeg hole, but mucho frio! You know—beeg cold air come out. Think she deep sonofabeech!"

This was too good to ignore. We pressed him for more detailed directions. He pointed to a large pine standing boldly atop the ridge a couple of canyons over.

"You go pretty much direction that tree. You cross small canyon, then top out. Cross next canyon and start up hill toward tree. Beeg bench. Hole in back of bench, down in hollow against rock wall. Plenty oak brush. Almost can't see hole 'til you fall in!"

We double-checked his directions, and watched him as he whistled Perro from the cool shade of the rock and headed off in the general direction of Black Cave.

He turned and waved and called out, "You find that damn hole, you be plenty careful! Lots of no good in them damn holes!"

We lolled about a while longer, reluctant to leave the cool shade.

"Well, we're getting nowhere fast," Dave grunted as he ground to his feet. "Let's go find the peon's beeg hole."

The sun had not cooled any. We got across the first canyon and to the top of the next ridge. Far off to the right we could see the look-out tower—it was still a long hot trudge to the car.

"You guys, I've had it," Ted exclaimed as he slumped onto a boulder. "I don't care if there's another Carlsbad Caverns over there, the only direction I'm going now is toward the car!"

The others agreed that some cooler time in the future would be better for hunting a cave that might be "over on the next ridge" and then again might just not. But a certain feeling had arisen within me, a feeling that this was right, that the cave was there, and that we would find it.

"It's not far over there. Let's go see. Besides, we can angle back toward the car without going too much out of our way."

"Nothing doing! If you want to go find the goddam cave, you go on. We'll wait here for you, and when you get to the ridge top, wave us on and we'll meet you at the car."

So I left them stretched out in the shade of a low pine and set off down the slope toward the floor of the canyon, keeping the Mexican's pine tree straight ahead of me. I scrambled through the rough boulders and oak shrubbery in the floor of the canyon and started up the opposite ridge. About one-third of the way up there was a large open flat area which sloped gently toward the mountain, and a ledge of rocks deeply buried in oak brush. And straight as a string, almost as though I had known it was there all the time, I walked right up to the mouth of the cave. I stood panting in the cold air pouring from the hole and could not believe my luck.

I called to the others that I had found the cave. I think they were sorry to hear the news. But they headed in my direction, and soon we were all standing in speculation before the opening. I still could not believe my luck; had I wandered only twenty feet on either side I would have missed it in the heavy brush. The hole was not large, about three feet wide and maybe five high, opening at the base of a small vertical ledge in a circular sunken area.

"Looks pretty good," Dave agreed. "Let's give her a whirl and see where she goes."

The corridor beyond the entrance was high enough to stand in, and sloped off rather steeply into the mountain. A short distance beyond daylight's penetration the corridor floor began breaking up into a series of short perpendicular falls dropping some thirty feet to the floor below.

"Doesn't look too good," Julian grumbled. "I'd better tie the rope around this knob, just in case." He secured the rope and shook it off into the crevice, but we found we really did not need it—by careful scrambling we reached the passage floor below. Smooth patches of tan colored flowstone had begun to decorate the walls, and a few squat, knobby stalagmites perched about. And it was cold!

The corridor ran smoothly off from the crevice, leading into the mountain, and gradually becoming smaller as we crept along. Finally we were forced to stoop to make our way, and then found ourselves on hands and knees, crawling along with the lowering roof forcing us ever lower. Dave was leading.

Suddenly he stopped, and we banked up behind him. We shone our lights around him and peered over his shoulder. The corridor ahead widened into a small, low-ceilinged room about the size of Grandma's kitchen pantry. And a few feet into this room the passage dropped into black nothingness.

As we gathered around Dave, he picked a large stone from the floor and pitched it into the void. A long silence followed, accentuated by the utter quiet; then the stone struck bottom, bounded twice, sending up rumbling echoes from below, and splashed heavily into water.

"Whoa, boys!" Dave shouted. "This is as far as we go! There's Hell below!" Thus was the cave christened, for although we gave it another name, whenever we found occasion thereafter to speak of it, we always referred to it as Hell Below.

With our lights we discovered that a huge chamber lay beneath us. The tunnel down which we had come dropped abruptly through the center of the ceiling of the room below. All back beneath us was hollow and vast, and our lights showed the big chamber continuing into the mountain farther than their feeble beams would penetrate. A cold draft swirled up about us, ample proof that the cave was huge and contained much water.

"Anybody for a trip below?" Julian joked. Our only lariat was back in the crevice, and, besides, it would take much more than one to reach the distant floor below us. We tied a light to the end of some strong line—which we always carried on a fishing reel to mark our way back through diverging corridors in any strange cave we explored —and lowered it to the bottom of the abyss, making a mark on the line when the light came to rest. A later measurement of this revealed that the drop from our perch at the end of the corridor to the floor of the chamber beneath was 68 feet.

"Well, we gonna stay here all day?" somebody grouched. "It's still a long way to the car. Let's get the hell outa here!"

Most of the way back to town we talked about the cave, and made plans to go back and explore it. But it was a month before we could get another trip organized. We stopped at the Shattuck ranch atop the mountains and several miles west of the cave, picked up horses and Julian, and rode over to the cave instead of driving the godawful road to Lookout Tower and walking over. Tommie Futch had joined us this time for the conquest of Hell Below.

The sun was not as hot now, and a fair breeze whipped across the ridges. Autumn was just around the corner. The cold draft of air still poured from the cave entrance.

"Remember how good it felt last time?" Julian exclaimed, as he shivered inside the entrance, adjusting his light.

We carried three lariats securely knotted together this time, extra flashlights and batteries, and lunches strapped to our belts. We were soon at the lip of the abyss. It had seemed a cinch on the drive up the mountain and the ride to the cave. We would make the drop in a breeze and be intrepid cave explorers trodding underground wonders never before seen by man! The view into the blackness beneath caused us to reconsider. I could feel our enthusiasm waning.

But we tied the connected lariats around a stalagmite about a foot thick that grew from the floor of the corridor and joined the ceiling five feet above to become a solid column, and tossed the line into the chasm. Our flashlights revealed plenty of extra footage lying on the floor below. Silence followed.

"O.K., who's gonna be first?" Dave asked. No one seemed in a hurry. Tommie finally grabbed the rope, straddled it, his back to the chasm, leaned back and shone his light down. He stood thus for many seconds, looking down, saying nothing. Then he stepped back to the column, carefully shook the rope clear, came back to us, and sat down.

"You know," he said in his slow Louisiana drawl, "I just don't think it'd be right for me to be first. You guys found the cave, and I don't want to steal any glory. So one of you lead off."

No one felt inclined.

We ate our lunches, and called each other several different kinds of sissies. While the others pulled the rope from the pit and coiled it, I found a flat ledge that to some extent overlooked the chasm beneath, poured out a good charge of flash powder on a newspaper fuse, opened my camera on time exposure, and fired the powder by lighting the paper. In the instant of the blinding flash we glimpsed the greatness of the chamber beneath us and the beauty that abounded there.

We followed the smoke from the flash powder as it rode the cold draft out of the depths.

"You know," Ted muttered, as we headed for our horses tied by the pine tree on top of the ridge, "those ropes tied together probably wouldn't have been strong enough. They might even have been dangerous."

Shades of orange, yellow, and red were touching the oak shrubbery, hinting at winter to come, when, a month later—on October 1, to be exact—we again gathered at the black opening of Hell Below. In the month interim we had screwed up our courage to a point of determination to plumb the depths. In preparation, Tommie had procured some 150 feet of half-inch manila stake line from the National Guard. We had doubled this and tied knots every three feet to aid us—we thought—on the ascent out.

Again we had stopped at the Shattuck ranch to get horses and have Julian join us. He swore this was the last time, goldang it, he was going with us to that so-and-so cave, and this time we'd better go all the way! We had also stopped at the Dave McCollum ranch and picked up Seth, Dave's son, who was somewhat curious about what these "damned fools" had found.

Back in Carlsbad I had maintained that, since I had been the first to find the cave, it was my right to be the first down the rope. At the edge of the drop, looking down into the darkness, I had reason to regret my stand. We tied the knotted manila securely around the column and dropped it over into the pit. Our lights revealed it lacked some fifteen feet of touching the floor. So we hauled it out and tied Julian's ever-present lariat to the end and pitched it back. Someone thought it would be a good idea if there were a light at the bottom to help us on our way down, so we lowered an electric lantern by separate line. It shone feebly far below, revealing nothing.

Then quiet closed in, broken only by the echo of dripping water. The others looked at me. I looked off into the pit.

"Well . . . ?" mouthed one of them.

I took a deep breath.

"Well, we came to see this cave, didn't we?" I said. "You guys stand back and let me on my way." No one was blocking the way down.

For about the first ten feet the rope lay flat against the perpendicular lip of the overhang. I grabbed the rope and backed down this, leaning out, feet against the wall. Then the line swung free and my feet touched nothing. I was into the pit. Wrapping my legs around the rope, I started down hand over hand toward the light. I realized almost immediately that the knots at three-foot intervals were a mistake. Each one meant a loosening of the legs to pass over, then a re-gripping of the legs and a loosening of the hands, ad infinitum.

By the time I reached the lariat tied to the end of the knotted stake line, the horrible truth was upon me: I could never climb out. When I eased over the last knot and onto the smooth lariat, my hands were so weary my grip was gone. I just could not hold the lariat tight enough. Gathering speed, I whizzed the last ten feet like a plunging meteor straight into the electric lantern, knocking it winding and the switch to "off."

I was lying on my back in awesome blackness, the echoes of my arrival thundering about me. As they died to a whisper above, I lay there, afraid to move, wondering how many bones were broken, whether I lay on the edge of another precipice, whether the floor beneath me were solid, whether a deep pool of water awaited my first move.

Then, in the silence following my crash, Dave's voice came down, quivering with anxiety, "Are you all right, Sam?"

"I think so," I yelled back. I realized, half-ashamed, that my voice was not as steady as usual. "Wait 'til I get my flashlight out of my pocket and I'll see." Few things have looked as good as that white ray of light when I snapped the switch. I was on good footing and in no danger. I found the electric lantern a dozen feet away, upside down but unbroken. Things looked better under its light. But when I looked up, the rope seemed to dwindle to a mere string before disappearing into the hole far above in the darkness.

"Everything all right down there?"

"Yeah, I'm O.K. But listen, fellows, I'll never be able to climb out of here, and one down here is enough; so don't any of the rest of you try it."

"You enjoy being down there by yourself?" Dave called down.

"I wouldn't exactly call it habit forming!" I yelled back.

And then I saw the rope swinging, and Dave came bumping down over the knots out of the darkness. His descent down the lariat was not quite as undignified as mine. Soon he was at my side, staggering free of the rope and peering up.

"Sonofabitch!" he exploded. "I'll never get out of this bastard!"

We were flashing our lights about, taking stock of our situation, when we heard puffing, a grunt, and a tortured "Damn" from above, and Ted landed with a thump beside us. It did not take much persuasion from us to convince the three still up on the lip of the drop that they had better not come down. After a short discussion they decided to leave Tommie above us to keep in touch while the other two rode back to Seth's ranch for a lot of rope and help.

"While we're here, we may as well see what we've got," Dave said. "It's going to be a long wait, and it's sure as hell cold standing around." About thirty feet from where we landed the cave pitched off at a steep angle, and the slope was slick, covered with seeping water. A pool gleamed at the bottom, and our powerful electric lanterns revealed only blackness beyond, with here and there a ghostly stalagmite standing sentinel.

"We might make it down," Ted said, hopefully, shining his light along the far edge.

"We might could have a bath!" Dave scoffed.

To our left the cave continued in a smaller corridor some twenty feet high. We started out along it, making our way over the rough floor. The walls from floor to ceiling were one mass of cascading flowstone, in deep rich tones of tans and browns. In perhaps a hundred yards the floor dropped off into blackness. Our lights revealed the floor continuing on some thirty feet below, but the drop was vertical. Rope would be needed here, and that was something we had none of.

The great calcite waterfalls amazed us. I tried to set up for a flash picture to record their beauty, but by now my powder had absorbed so much moisture it would not fire, and all I could hope for was a long time exposure by lantern light. Rocks and ledges and protuberances bore à coating of richly colored calcite, resembling huge cakes covered with caramel icing.

"Just like Aunt Lucy used to bake," Dave muttered, which reminded us we were hungry, and our lunches were back at the foot of the rope. After eating some of our sandwiches—saving some "just in case . . ."—we decided things definitely looked better.

"I'm not real sure I can't climb out of this hole," I exclaimed, looking up the rope. "I think I'll give it a try." I managed to sputter and grunt my way up to the junction of the lariat and the knotted stake line, and realized what a dreamer I was. But my descent was not as precipitous as the first time.

"Hey, here's a skeleton," Dave called. It was almost at the foot of the rope, but we had overlooked it in the excitement of arrival. It appeared to be either a coyote or a fox, and was now completely encased in limestone, an integral part of the cave. Closer search revealed the tiny bones and skulls of innumerable rodents scattered about, all deeply encrusted and firmly cemented to the floor where they had fallen from the corridor far above.

"Hey, looky!" Ted exclaimed. "Pearls!" He had found a bed of cave marbles just a few feet from the coyote skeleton. There were dozens of them, each lying in its individual cup where it had formed. We had never heard of cave pearls larger than a child's small marble, and yet we found many here much larger. I picked up one round and smooth and big as a billiard ball. Some glistened smoothly, as though polished on a lapidary's wheel, while others were rough-surfaced, resembling the moon as seen through binoculars.

I remembered a cave scientist explaining that these formed when drops of water, heavily burdened with limestone in solution, fell from a great height and rounded out a small depression where they struck, and in so doing caught a small particle, such as a grain of sand. As the drops of water built up limestone around the grain and it grew and grew over the numberless years, it was kept round by the constant turning under the force of the falling drops. And although the falling water also built up the surrounding floor area with a deposit of calcite, the turning of the growing marble kept forming a cup which grew in size corresponding to the growth of the marble.

We had about exhausted the things to be seen, and there still was no sound from above, except an occasional shout from Tommie, keeping vigil on the ledge. Occasionally he would exclaim, "I'm freezing!" and head for the entrance to warm up in the sunshine.

I decided to make a try for a picture of our rope going up into the darkness, so set my camera on its back on the floor. Dave had a small bottle of flash powder that he had not opened; hopefully, it would still be dry in all the humidity of the cave. I pulled a dry sheet of paper from the inside of my roll, hurriedly rolled it into a fuse and had Dave pour a charge of powder onto the end while I clicked the camera shutter onto time exposure and lit the paper fuse. The flash powder sputtered a second, then went off with a dull boom, and I knew I had some sort of picture. When I snapped the shutter to "close" I realized it had not clicked, so I hurriedly wound to the next exposure, keeping the lens covered. Sure enough, the shutter was stuck on "open." The humidity had got to it.

"I guess that ends our picture taking," I muttered. However, I did try a few exposures by lantern light simply by uncovering and covering the lens.

Our teeth were soon chattering from the cold, and we jumped about to keep warm. The extreme humidity of the cave saturated our clothing, adding to our discomfort.

"I think we misnamed this hole," Dave said. "Hell was never like this."

I remembered what the Mexican goatherd had said: "Lots of no good in them damn holes!"

We spent several uncomfortable hours before we heard voices echoing from above, and knew our rescuers had finally come. They had brought Seth's father and three thirty-foot lariats.

"You guys about ready to come out of there?" Seth called down.

We assured him that was a gross minimizing of our thoughts. They knotted the three lariats together, and soon the end came snaking down beside the knotted stake line.

"One of you tie the rope around you under your arms," Julian called down. "Be sure you tie it tight, and we'll haul you up as you climb the other rope. Don't worry, we'll get you out."

"Well, Sam," Dave drawled, "since you outweigh either of us by several pounds, you go up first, and if everything holds you, we know we can make it."

"Thanks, pal," I muttered as I tied the rope under my arms and across my chest. Dave stepped up and checked the knots to be sure they were tight. Then he touched my shoulder and said softly, "You'll make it all right, Sam."

I admit my heart was not exactly in my work as I started up, but the steady pull of the supporting rope was a comfort. All went well for the first thirty feet; then I started twisting around in spite of all I could do, and the two ropes became more and more entangled with each other. Just a few feet short of the bottom of the ledge I could go no farther. The knots we had tied in the stake line only made matters worse. So I hung there like a pendulum in space, frightened, untangling the two ropes. I finally got them sufficiently cleared that I could go on.

As I got to the bottom of the ledge, where my weight pulled the stake line taut against the solid rock, the thought went through my mind, "Just how in the hell did you think you could get past this climbing out on your own?"

But I finally reached an outstretched hand, and was dragged up flat on my belly to the safety of the ledge. I just lay there and panted. I think I must have patted the ground. Nothing had ever felt so good.

In a little while they yanked Dave up to the spot I had vacated, and I heard him exclaim, "Sonofabitch, am I glad to be out of there!"

There was no trouble at all getting Ted out, as he was much the lightest of the three. Then we pulled up stakes and headed for the surface. I have never felt sunshine so welcome, nor seen it so bright.

5
Falling Rock Cave

NEVER AGAIN DO I WANT TO HEAR that song "The Death of Floyd Collins," for it brings back memories of a lost cave in the Guadalupes, and I see again Julian's blanched face in the dim light of our lanterns, and hear the rock rolling down the corridor toward us. And Dave's ridiculous whistle . . .

We were up to our old tricks that morning. Dave Wilson, Julian Shattuck, and I, hunting new caves to explore. We found the opening to this one dropping straight down for about twenty feet in the rugged rim of a small canyon on top of the mountain a few miles northeast of the old Queen townsite. This hole lay several miles north of the most productive cave country, and we doubted it would amount to much, but the old rancher who had told us how to find it had said, "I betcha it goes clean to Hell."

"I don't like the looks of it," I muttered as we peered down into the blackness. The rocks about the entrance seemed loose and insecure. The shaft yawned black and uninviting in the shade of a huge juniper clinging precariously to the canyon's rim. However, Dave tied our lariat around the base of the tree, and we lowered our heavy pack—containing food, camera, flash powder, string, first-aid kit, and batteries—to the bottom, and began clambering down.

I noticed as I slid down the rope that the narrow sides of the cleft seemed jagged and broken, as though the rocks had cracked under tremendous pressure. Debris that had fallen from the walls covered the bottom of the shaft. Strangely enough, the bleached skull of a bull—Julian thought it might be a buffalo—mouldered there.

"For a minute I thought it might be my Uncle Mortimer," Dave commented dryly as he stepped over it and turned into a low arch leading off into darkness. We had to stoop to get under, but then progressed upright into a small corridor that sloped down at a sharp angle, winding to the right beyond the beams of our lights. As we followed the tunnel, it continued down, winding always to the right. Like a huge corkscrew it bored down into the mountain.

The going became rougher. Huge boulders littered the floor or hung precariously from the sides of the tunnel. In places the ceiling dimly reflected our lights thirty feet overhead, but the distance from ceiling to floor averaged about ten.

"Sure isn't much to look at," Dave grunted as he pulled himself up over a broken ledge. The barren brown walls bore no cave decorations; then, as we advanced, unimposing stalactites began to appear. Small, brown, expanding at the tips, they resembled miniature spears. Dry and unlovely, they seemed brittle to the touch.

The tunnel continued down into the earth. At times we had to lower ourselves over ledges and piles of huge rocks. Occasionally I felt the boulders tremble as I shifted my weight across them. When we talked, the echoes broke hollowly, as though the walls swallowed up the tones.

"I've seen better caves," Julian panted as he stopped for a breather.

We wound down and down. It became hot—almost stifling. No fresh air circulated here, as it had done in most other caves we had explored. Small, straggly, dead stalactites still appeared intermittently, or narrow, dry, paper-thin sheets of onyx wound across the ceiling where moisture had seeped from the cracks. Jagged-edged and sharp, they reminded us of long rows of shark's teeth.

We came at last to a point where a pile of loose boulders half filled the tunnel. As I reached up to pull myself over the obstruction, a large rock came loose in my grasp. The crash that should have followed hardly echoed as the walls absorbed the noise. I stamped my foot. Only a soft thud resulted, but we could feel the ground underfoot reverberate hollowly. Another rock came loose from the side of the corridor and rolled toward us.

"Jeez . . . !" I heard Julian draw out with bated breath. Suddenly we realized that all the walls about us were rotten, that the cave we traversed was crumbling, and that what lay ahead was ready to collapse at the slightest excuse.

We looked at each other, and Julian's face gleamed white in the darkness. I could hear the others breathing. Then Dave, whose sense of humor, although slightly warped at times, never left him, began softly whistling "The Death of Floyd Collins."

"Shut up that damned whistling," Julian growled. "Or if you must, pick a different tune."

We started cautiously retracing our steps, all interest in what might lie ahead gone. I have no idea how far down into the earth we had penetrated, but certainly several hundred yards separated us from daylight. The way back seemed interminable. Great boulders quaked uneasily beneath me as I clambered over them, and I held my breath, expecting each moment to hear the walls come crashing in. The heat of the corridor, abetted by the tenseness of our situation, soon soaked our clothing with perspiration.

"This is supposed to be FUN," Dave exclaimed as he inched carefully up and over an unstable boulder. I think we all breathed a sigh of relief when we finally emerged from beneath the low arch and saw the white skull gleaming on the floor.

"Uncle Mortimer," Dave drawled, "you never looked so good."

He went up the rope first. "Okay, all clear," he called down. I tied the pack to the end of the rope, and Dave started pulling it up. It hung on an obstruction just out of reach, and I stepped up on a small ledge to free it. As I reached the ledge, Dave yanked, and the side of the shaft gave way. I saw the rock coming, and ducked my head just as it struck me.

Julian was helping me to my feet from the bottom of the shaft. Dust filled the air, choking us. I brought my hand away from my head, sticky with blood. I could feel the stuff running down my face.

"What in hell's happening down there?" Dave shouted.

"Robert's hurt—head's bleeding pretty bad," Julian called back. "I'll help him up the rope and you grab his hand when you can reach it. He's pretty shaky."

Dave pulled the pack on out and lowered the rope back to us. Julian helped me back to the ledge, and I started climbing the rope. I was about halfway up when a wave of nausea hit me, and I reached

Falling Rock Cave 63

out to some protruding rocks to steady myself. Julian balanced directly below to lend a hand. I had hardly put my weight on the jagged wall when the entire mass crumpled beneath my hand. I almost tumbled with the wreckage to the bottom of the shaft.

"Julian . . . ?" I called out when the noise of the crash had subsided, but he did not answer. Then I heard him groan. Sliding back to the bottom, I found him just rising to his knees. Blood covered his head. It streamed down over his shirt and spattered on the rocks. I forgot my injury when I saw how badly he was hurt.

"What is going on down there?" Dave yelled down.

"Julian really got it," I called back.

"Do you want me to come down and help?"

"No, but be ready to grab him when we get there."

I started assisting Julian up the rope, carefully testing the sides of the shaft as we mounted. He was so groggy I was afraid he might slump over against the wall and the whole thing would come down. Luckily, no more rocks broke loose.

"My God . . . !" Dave exclaimed as Julian's bloody head poked into daylight. Tense and white of face, he was not whistling a forlorn melody now.

When we got Julian into town some four hours later, the doctor took over a dozen stitches in his scalp.

"It's a good thing you two fools were wearing your sombreros," he muttered, "or it could have been a different story. Cave exploring! Humph . . . "

Julian and I still carry three-cornered scars on top of our heads where the falling rocks struck us, which explains the name we gave the hole—Falling Rock Cave.

And that's why I don't like the song "The Death of Floyd Collins."

6
Hidden Cave

IN THE SUMMER OF 1926 I went to work for Ray V. Davis, the pioneer Carlsbad photographer who did the first extensive photography of Carlsbad Caverns. I was a senior in high school when the fall term opened, and continued working for Davis after school hours and on weekends. My first duties consisted of taking finished prints from the washer, placing them face down on black enameled "squeegee" plates, running these through a clothes wringer, and watching over them until they became dry and popped off the plates, slick and glossy. Many of these were prints of Carlsbad Caverns, which Davis sold in his studio.

But one afternoon when I reported for work after school, I started drying some prints of a beautiful cave that I knew was not Carlsbad Caverns.

"Those are in Hidden Cave," Ray replied to my inquiry.

"Where's that?"

"It's a cave up on top of the mountains, close to Dark Canyon Lookout. I made the pictures for Huling Ussery to publicize his dude ranch and mountain horseback trip. He takes his dudes into the cave as one of the features of his mountain ride."

"It sure looks like a pretty cave," I remarked, removing some prints from the washer. "I think this picture here is just as pretty as anything you have of the Caverns."

"It's a pretty cave, all right, but doesn't approach the Caverns in size. It's a real wild trip through it, and I'll bet some of his dudes give him trouble."

It would be more than seven years before I actually got to see Hidden Cave. By then the cave bug had bit me in Spider Cave, and I had several cave exploration trips behind me. It was, in fact, on our way back to the McCollum ranch after Seth and his father had helped get Dave and Ted and me out of Hell Below. As our horses topped a short rise, Seth pointed across an intervening valley.

"Right over there is Hidden Cave," he said. "See that clear shoulder about halfway up, right in front of a grove of pines, with what looks like a small stack of logs? That stack is a log gate the Forest Service has built over the entrance. They keep it locked so people won't get in and break up the cave. Dad keeps a key; someday we'll go over and have a look."

The day came on February 4, 1934. Julian Shattuck and I were not feeling our best. The day before, with Dave Wilson, we had met catastrophe in a cave we named Falling Rock Cave, when the entrance-drop collapsed as we were climbing out, and our heads took the brunt of the fall-down. Dave was fortunate enough to be on the surface, and helped us out. But when Seth asked us if we were up to a trip into Hidden Cave on the morrow, we forgot throbbing skulls and jumped at the chance. I still remembered those beautiful photographs I had seen in the Davis studio.

The four of us pulled up in front of the barricade over the cave's entrance in the early morning.

"Does Huling Ussery still bring dudes to see the cave?" I asked Seth, as he fumbled with the lock and chain, swearing at its stubborn resistance.

"I don't think so," he answered. "Not many, anyway. I think his dude ranch operation has about played out, with the depression and all."

He got the gate swung back out of the way, and I saw a small opening about three feet square dropping down between some large boulders. A makeshift ladder led down into the darkness, and I do mean makeshift. Someone had taken the lower fifteen feet of a pine tree, lopped off the branches about six inches out from the trunk, and dropped it through the entrance to the floor below. No one could call it neat, but neither could anyone deny its effectiveness.

The narrow passage we found ourselves in expanded after a few steps into a large, low-ceilinged room about fifty feet across. Many delicate white stalactites hung from the ceiling, and we could see the stubs of many more; we had to stoop to miss those remaining, and knew that tall men in tall sombreros had accounted for the others.

A small corridor branched off almost immediately to the right. We could span it with outstretched arms, but the ceiling gave plenty of head room above. Within a few yards the passage ended in a slightly wider room with a much higher ceiling. A pool of water covered the floor from wall to wall, and these walls were covered with rippling brown flowstone, wet and shiny.

"Take a look at that ceiling," Dave cried out.

It certainly was the room's crowning glory. Hundreds of stalactites in every conceivable shape hung there; we saw soda straws, helictites, elephant ears, draperies, coral clumps, grape clusters, strands of popcorn. But best of all were the two or three dozen smooth ones about wrist size and three feet long that ended in smooth round or elongated knobs up to six inches in diameter.

"If those don't look like Indian war clubs, I'll eat my hat," Julian remarked.

Seth told us the room ended just around a corner ahead, but even had it gone on, we could not have crossed the pool. So we returned to the main section, turned right, and soon entered a much larger room, with a ceiling twenty to thirty feet overhead. We had never imagined a cave with so many decorations. As we wandered along we were continually amazed by their numbers and variety. In many spots stalactites completely blocked out the ceiling; they ranged from tiny ones the size of macaroni to big fellows ten feet long and as big as my leg. Many joined the floor to become columns. Pure white mingled with soft tans and chocolate browns.

Stalagmites completed the picture. They ranged from the small short ones, like tent pegs, standing everywhere, through those of hitching-post size, up to huge monoliths a dozen or more feet thick at the base and tapering to a point fifteen feet above. In a small side chamber we found a beautiful column fifteen feet tall, its sides deeply fluted and its entire surface rough, like coral heads. It leaned slightly out over a crevice whose walls bore the same rough coating, as did the chamber's walls and the hundreds of stalactites that hung from above. Dave suggested we call it the Leaning Tower. I do not really know if he had in mind the famous structure at Pisa.

"Have you noticed how so many of the formations look like coral?" I asked. "They're like great undersea gardens."

We stooped into a small side room not more than ten by thirty feet. The low ceiling was lost in a forest of white stalactites hanging down as thick as reeds along a river bank. Most of them sparkled with dripping water, and they ranged in size from a pencil to a man's arm. Many slender columns joined floor to ceiling.

"How about the Icicle Room for this one?" I suggested.

After two or three hours of wandering in the many rooms and passages of this section of the cave, we started down a corridor that stretched straight ahead.

"Hey," Dave shouted, pointing ahead, "that's daylight. We couldn't be back at the entrance."

He was right on both counts. We soon walked under a natural skylight, a long, narrow gash high in the rocks overhead. It seemed to be about five feet wide by fifteen feet long, and some thirty feet above us. And there was a drop-off before us; we could see the cave continuing in the daylight streaming in from above. There was another pine tree ladder here, quite a bit longer than the first one, and by climbing down it we reached the lower cave.

Just as we made a sharp right turn and left daylight behind, we entered a chamber much larger than any we had seen thus far. Our attention was immediately attracted by what had to be this cave's most unusual sight. I think the same thought came to all of us.

"The Wall of China!" someone exclaimed.

Starting about midway against the base of the left side, a miniature winding wall, almost uniform to a height of about eight inches, snaked out, followed the wall for several feet, then wound out into the center of the chamber and circled toward its far end, zigzagged in graceful loops almost to the far side, then circled back to rejoin the left wall twenty or more feet from where it started. This amazing structure maintained strictly vertical sides and a uniform thickness of about half an inch.

The wall enclosed a natural replica of a Chinese garden in miniature. Near the front of the garden stood a perfect, low, broad-based temple, rising in five terraces to a flat and spacious top. Surrounding the temple was a level field; then a forest started and continued down to the far curves of the wall. The trees were tiny rough stalagmites about four inches high. One could even imagine miniature bridle paths winding in and out among the trees. Outside the wall, the floor of the big room was almost bare. Its walls, however, were decorated in glistening brown flowstone, and at the far end of the room, beyond the Chinese Wall, stood a teepee in stone, brown and shining.

We continued into the cavern, which remained large and beautifully decorated. The ceiling was dim in our lights; we judged it to be forty feet overhead. In many places the walls were sixty to seventy feet apart. We wound in and out among great stalagmites on a smooth and level floor. Long lines of them, solidly joined and turreted, towered twenty to thirty feet in the air and divided the room like great partitions that never quite reached the ceiling.

We explored several small side rooms that were marvels of serene beauty. We made our way down into one that was a dozen feet beneath the main level. It was rather narrow, about ten feet across, but about twenty feet long, with the ceiling two dozen feet overhead. This chamber had once been the site of an underground lake, for the walls were covered with great spongy-looking masses of limestone resembling huge clusters of cauliflower. The floor was level and smooth, and only a few stalactites decorated the ceiling. Our voices echoed flat and hollow. While setting up for a picture I accidentally knocked against a wall, and a large mass weighing many pounds crumbled loose and crashed to the floor.

"Jesus," Julian commented, "this place is as bad as that hole we were in yesterday."

We found that only a heavy tap was needed to bring huge sections of the cauliflower wall tumbling down. I hurriedly finished setting up for the photo. None of the others would get in the picture, for fear the reverberations of the flash might bury them. However, everything held when the powder flared. Dave suggested we call it the Coral Room Looking for a Catastrophe.

At one spot we found our way blocked by a high wall of limestone

46

47

48

whose top still lacked several feet of touching the ceiling thirty feet overhead. Seth took the lasso he carried around his shoulders, and, on the second try, neatly settled his loop over a squat thick stalagmite perched atop the wall. With the aid of the rope, we wall-walked our way up and over into new chambers of infinite beauty. Water dripped everywhere from hundreds of stalactites and seeped down over walls plated with rippling brown flowstone.

49   We eased down through a narrow opening into a fantastic room we named the Lost Chamber. This, again, was the site of an ancient underground lake, now dry, with only a small puddle in one corner. Great masses of what we called coral heads hung out from the walls. Some were taller than a man and four feet thick. Long columns of them hung from ceiling to floor, resembling strings of sponges hanging up to dry. Everything was a dull, lusterless brown. Small, jagged stalagmites covered the floor, resembling an evergreen forest as seen from a great height. Unlike the other room, where everything crumbled to a touch, this room was stable, with the decorations firmly anchored and solid.

"Fill this room with water and add a few fish," Julian remarked, "and you'd have the coral gardens." It was a room of rare beauty.

We started a long hard climb up a steep corridor. After a score of yards it began to pinch down, and then ended abruptly. At the upper end, fifteen feet overhead, we saw a cluster of tiny white stalactites just too long and too lacy to be true. We clambered up a shaky break-down to inspect them more closely, and found they were not stalactites at all; they were delicate hair-like roots of plants that had grown down into the cavity, and subsequent dripping of water had coated them with paper-thin sheets of calcite. Not many feet separated us from the outside surface.

We backtracked to the main cavern and continued. We soon found our way blocked again by a dividing wall, and put Julian's lariat to work to get over. The cave was narrower beyond, becoming a corridor about fifteen feet square filled with stalagmites glistening wetly in our lights; many pools of water, whose surfaces were constantly dimpled by falling drops, filled the sunken areas. Occasional circular pedestals, corrugated and flat-topped, rose above the water, like ancient fountains. Many stalactites, some broad at the tip like spears, hung down amid rows of draperies and elephant ears. Everywhere were the inimitable frozen cascades of brown flowstone covering the walls.

"We've got to call this the Fairies' Palace," Dave said.

50

"Suits me fine," Julian retorted, "just so you let me eat." He hopped across a small pool to a projecting ledge of pale tan flowstone, plopped down, and dragged out a sandwich.

"This may be a fairy's palace," I remarked, "but he looks just like a little mean gnome."

"Yeah," Dave came back, "or a big, fat Louisiana frog on a lily pad."

Seth decided a spot on the other side looked more comfortable and hopped across an arm of a pool to a fountain. He then leaped for the opposite edge, lost his footing and his balance, and to avoid falling over backwards stepped back into the water, going in to his waist.

His yowl of dismay rent the welkin. He scrambled out onto the rough ledge, gasping and swearing.

"Kinda chilly, was it, Seth?" Julian asked dryly.

"Chilly, hell, that was just plain damned cold."

While Seth pulled off his water-logged boots and wrung out his socks, we finished our lunch, and I set up for a picture. I found a ledge high up above the pools where I could place my camera, and then laid out three separate paper-triggered flash charges: one on the floor beneath the camera, which I would light; one twenty feet beyond in a side tunnel, which would be Dave's responsibility; and the last one fully fifty feet away around a bend, which Julian would take care of. With shutter open and matches lit, I called, "Everyone light paper." It always amazed me how nearly together these "synchronized" flashes would flare up.

Seth, grunting, put on his wet boots, and we went on, following a long, winding corridor that was never more than eight feet wide, but often the ceiling soared thirty feet overhead. Generally, the floor was smooth and level, but the walls disappeared behind masses of limestone waterfalls and pleated draperies. Many were the color of flowing taffy. A few stalactites appeared here and there on the ceiling, but this corridor, which we called the Grand Canyon, owed its beauty to the magnificent decorations covering its walls.

"Fellows," Julian remarked, as the Grand Canyon pinched down, giving a promise of arduous exploration, or none at all, "we've got to be getting out of here. It's going to be long dark by the time we make it to the ranch."

"I won't argue with you there," I replied, suddenly realizing that weariness had already started creeping into my bones and muscles.

On the way out, and long before reaching the skylight entrance overhead, Seth led us over to a protruding wall and pointed out some large bones partially buried beneath the overhang. We scratched around in them, uncovering a part of one larger around than my thigh. Although we could not locate the skull, we estimated what remained of the skeleton to be about eight feet long.

"What the hell was it?" Dave asked incredulously.

"Some guy that knows bones thought it might be an extinct cave bear," Seth answered. "He sent a couple of the bones to a museum, but I don't know if he ever heard from them."

We reached the pine tree ladder leading up to the first level, just beyond the Chinese Wall Room; we could see the faint light coming in through the natural crack up above and ahead of us.

"You know," Seth began, "Uncle John told me that one time when Huling was bringing a bunch of his trail riders through here, they ran onto three big rattlesnakes right beneath that opening."

"Rattlesnakes!" Dave interrupted. "How could rattlesnakes get down here, and why would they want to come here in the cold in the first place?"

"I don't know," Seth went on. "They probably got involved in their lovemaking outside and fell through that big crack up there. Anyway, there they were. I'll bet some of those eastern dudes damn near stampeded when those snakes let loose with their bells."

"Yes, and we'd better stampede the hell outa here," Julian growled. "Look at that daylight up there. Just about gone. It'll be dark by the time we get out of here."

And he was right. Only a whisper of daylight lingered over the far western ridges of Dark Canyon as we emerged from the cave, and pine trees skylighted on the scarp above us made only a suggestion of silhouette against the slightly lighter heavens.

We had spent more than eleven hours in Hidden Cave, and the muscle-grinding weariness of those hours became increasingly apparent as we started the long, rocky hike to the clear spot where we had parked the car.

Now we really began to ache. But, as Dave put it when he tumbled exhausted into the Model A's rumble seat, "I guess we gotta feel like this in order to realize the day hasn't been just a dream, after all."

7
Endless Cave

"I KNOW THIS IS McKittrick Cave," I said, "because I've been here before. And I'm positive this isn't the cave that guy was talking about." Dave Wilson, Tommie Futch, Glenn Hamblen, Sonnie Kindel, and I had braved a whistling sandstorm to look for a cave a man at a service station had told me about.

"It's on beyond McKittrick Cave," he had vaguely directed, "up kinda high on the hill. You'll have to walk maybe a mile to get to it. You can't miss it—there's an old broken-down wire corral around the entrance, where some old sheepherder used to pen his flock in bad weather. Just keep on circling around the north side of the draw and you'll find it."

We had parked our car beside the manhole entrance to a cave which they thought might be the one, but which I knew was McKittrick Cave. Besides, we had just about hit the end of the so-called road, and my directions had called for "maybe a mile" walk.

"Well," Tommie drawled, "we've got a cave here, and you know the old saying about a bird in the bush. Why not look into this one?"

"I've been in there, and there's nothing left to see," I replied. "People have been hauling formations out of that cave since Mother was a girl—bare walls and ceilings just don't appeal to me."

"Well," Glenn said, "let's just take that mile walk and see what turns up."

It was not much more than a mile before we stood in front of a black hole at the back of a shallow depression on the north slope of the hill, the peak of which was plainly in sight not more than a hundred feet up the slope. The entrance was an oblong hole, about four feet high and ten feet wide, the top of which appeared to be a solid capstone the entire length of the opening and about two feet thick. The old corral enclosing the entrance slumped tiredly, the barbed wire touched the ground in many places, and some posts were broken or missing.

"Sure enough looks interesting," Sonnie said, firing up his gasoline lantern. "I'm ready to give her a try."

We had to stoop to clear the capstone, and immediately found ourselves in a long, narrow entrance chamber about twenty yards wide and of undetermined length. Within a few feet we could stand upright, and noticed the cave branched, with a broad corridor about four feet high leading off to the right, while the main passage continued on ahead. Dave thought we ought to try the branching corridor, so we turned down it. It proved to be a disappointment, consisting of only one low-ceilinged room about thirty by sixty feet square, with no side passages.

We went back to and continued down the main passage. About forty yards from the entrance the ceiling dipped close to the floor, and we had to stoop low to negotiate the pass. In a few feet the rock vaulted overhead, and we found ourselves following a passage some eight by ten feet square. Dozens of large white stalactites hung down, and short, thick stalagmites squatted on the floor. They were white also, but we noticed a couple that had been broken off, and discovered that they were really almost black throughout, with just a thin coating of white calcite changing their appearance.

"Well, someone's been here before us," Dave muttered. "Those didn't get broken by bats."

We followed this small, twisting corridor for about a hundred feet; then suddenly it opened into a large room probably thirty feet square with a ceiling twenty feet overhead. The floor sloped gradually up to the center, forming a low, flat-topped mound in the top of which nestled a small pool of water about eight feet across and two feet deep. A paper-thin sheet of calcite had started building up across the water's surface from the edges, and covered the entire surface except for the center.

"Looks a little bit crummy," Tommie grumbled, "but I'm already thirsty." He brushed away some of the crust and leaned over and drank long and enjoyably. "Nothing wrong with that," he said, wiping his mouth. "Certainly have had lots worse down in the swamps of Louisiana."

We named the pool of water First Spring. The walls of the room were a dark brown, broken here and there with rippling tan flowstone; some tiny stalactites dangled from miniature ledges. The floor was covered almost entirely with a deposit of brown, cascading flowstone. Clustered around the edge of the pool stood a half dozen broad-based stalagmites, squat and brown, about a foot high. One towered over its companions, tapering to a sharp point four feet above the pool. Several dozen drab stalactites hung down, admiring their reflections in the water.

Glenn flashed his light down the corridor ahead. "Well," he said, "we seem to have two choices; which shall it be, left or right?"

A small tunnel led upgrade to the left, bending to the right out of sight, while the main corridor seemed to go straight ahead. We continued on down it, and within a hundred feet the floor dropped abruptly, forming a rough crevice about fifteen feet deep and ten wide. We were able to cling to a narrow ledge that skirted the left edge of the cleft. We came upon a pool of water, not more than four feet across, but of such a depth that the beams of our lights disappeared in the green water. I tied my knife to a spool of twine and lowered it into the pool. It was over eleven feet deep.

"If you're still thirsty, Tom, there's plenty here."

The crevice ended as abruptly as it had begun, and we continued on down the corridor. Soon, however, the passage dipped suddenly over a ragged drop of about fourteen feet. We found toeholds on the slippery side and eased ourselves to the bottom. Here the tunnel was about twelve feet wide, with the ceiling two dozen feet overhead. We could see a long row of brown stalactites about two feet long following a twisting crack in the ceiling, like a winding, inverted picket fence.

A deposit of firm mud, just wet enough to be sticky, covered the floor of the crevice. Near the base of the drop that we had just clambered down, a small, round hole, just big enough to squeeze into, branched to the left, like a rat hole in the white wall. It beckoned for exploration, but we decided to follow the main corridor, and soon had to crawl up over a rough wall much like the one we had just come down. The corridor had reverted to its original size.

We went on. The passage was very rough, with great rocks strewn over the floor where they had fallen from above. For a couple of hundred feet we struggled through the rubble, the corridor growing steadily larger. Finally it was over thirty feet wide. The ceiling, twenty feet above, was dim in the lantern light. Very few formations decorated the drabness of this passage, although scattered spots along the walls were covered with sheets of brown flowstone.

Then the corridor narrowed again, forming a crevice about fifteen feet deep and three feet wide, in the bottom of which water cast back the beams of our lights. At first glance it did not seem possible to go on, although we could see the corridor continuing where the crevice ended about twenty feet ahead. Tommie said he thought he could make it. The right edge of the pit had a small ledge about three inches wide its entire length, offering a foothold, and the opposite wall seemed irregular enough to give support. Tommie put both feet on the small ledge on the right and leaned over to the opposite wall with his hands and began crab-walking to the other end of the crevice. He seemed to make it with very little difficulty.

"Come on over," he called. "It's chicken pie."

Sonnie made it over without difficulty; then Dave started, with Glenn close behind. Dave had just made it to the far end and firm footing when Glenn's boots, wet and slick from the mud in the crevice behind us, slipped on the precarious foothold, and with a great clatter he fell into the pit. I thought he was a goner, but somehow he managed to hang on, with hands and elbows braced on almost nonexistent projections, his legs flailing for support. I can remember his white face gleaming in the pale light.

"How deep is that damn water?" he yelled.

"No need to try finding out now," Dave said, calm and cool. He had turned, with feet on firm footing, and leaned over within easy reach of Glenn. "Here, grab a hand." Glenn hesitated a moment, fearful of relinquishing the slightest hold, then grabbed wildly out with his right hand, and Dave snatched it. Fortunately, Glenn was not the heaviest of cave explorers, and Dave was pretty much of a man; he just heaved Glenn up to the solid floor of the corridor.

"Whew!" Glenn exclaimed. "All I could think of was that water down there and me not much of a swimmer."

"I don't know," I called across. "Maybe I'll just wait for you guys to come back."

"Who says we're coming back this way?"

I did not look forward to it, and I did not enjoy it, but I made it across without mishap, and we continued on up the corridor. After only a few yards we found a narrow, flat hole leading down at the base of the left wall. Suspended from the ceiling above was an oddly shaped stalactite about three feet long; it was thin and curved, resembling a scimitar.

"Let's try a side tunnel," Dave suggested. He slid into the hole feet first, and soon called to us to follow, as the passage seemed to lead somewhere. I was just able to squeeze through the small opening by extending my arms over my head, and was glad to find the passage enlarging immediately. It sloped sharply for about fifteen feet, opening into a small room, the floor of which was almost covered by a pool of water. The entrance slope and the floor of the room were completely paved in rippling flowstone, and the walls dropped in cascades of limestone of a deep raw sienna brown, glistening with water seeping everywhere.

A small opening, almost closed by stalactites, led to the right from the pool of water. We could see a room beyond, beautiful in the beams of our flashlight.

"Sure would like to get into that," I muttered.

"The only way we'll do that is break some of these out," Sonnie said, tapping a stalactite barring the way. It rang melodiously, like a chime, and the echoes came back from the forbidden room beyond.

"If it's the only way, then it's the only way," Dave said. He grasped one of the stalactites and snapped it off. Reluctantly we broke off about six of them, enough to give us a tight passage into the room beyond. Hesitant to break more, we sent Glenn, who was the smallest, through to reconnoiter.

"Sure is purty," he called back. Then his light grew dim and disappeared as he worked his way around a projecting wall. Not many minutes elapsed before he came back into view, shining his light about. "Room's almost circular," he reported as he squeezed back to join us. "Sure has lots of nice formations and some of the prettiest flowstone I think I ever saw."

"Is that all?"

"Well, there's a real interesting twenty-foot drop around at the far side. I could see a good-sized room leading off from the bottom. But it'll take some rope to get down to it—anyway, I'd want some rope before I'd try it."

We crossed the pool of water and followed the tunnel, having to stoop most of the way. For probably a hundred feet it wound about, then opened onto a shelf which overlooked a section of cave much larger than anything we had found so far. The corridor was about thirty feet high and half as wide, rugged and rough, with large rocks tumbled about on the floor. At one end of the room, just within range of our lights, stood a beautiful white stalagmite about ten feet high and eighteen inches in diameter. Tall, straight, and stately, it tapered gracefully in receding tiers, like a lighthouse guarding the canyon. With complete lack of originality we called it the Totem Pole. Three rows of tiny tan stalactites followed as many cracks in the ceiling, and several hefty stalactites hung there. Large flows of brown calcite decorated the walls, and small yellow and tan and white stalagmites perched on narrow ledges or the tops of fallen rocks.

55   "This is by far the grandest thing we've found so far!" I exclaimed. "Let's call it the Grand Canyon."

We scrambled to the floor of the canyon and started down its length. Straight as a javelin it went for about a hundred feet, and at its end we came upon the most beautiful mass of flowstone we had ever found. Starting up near the ceiling twenty feet above, in a deep chocolate brown mass it poured down over the wall, bulging here, drooping there to form little fluted cascades, finally to stream out over a ledge and pour down in beautiful, delicately thin elephant ears and folded curtains, the bottoms of which hung only a couple of feet above the

56   floor of the corridor. Behind the curtains was an alcove into which Dave stepped and held his light close against the draperies; they were translucent, and the light came through soft and golden. The entire thing resembled a great mass of molten taffy that had spilled over the edge of a pan and congealed as it dripped from the table.

"Taffy Hill," Sonnie called it.

Just behind this the Grand Canyon narrowed abruptly to a small passage through which we could barely pass, stooping low. A small corridor led down at an angle, its floor soft and sandy underfoot. Within thirty feet it opened into a circular room about fifty feet across with a twelve-foot ceiling. Long, slender stalactites, some over six feet long tapering to a sharp point, clustered overhead. Mostly they were brown, ranging from tan to a deep chocolate, but occasionally a pure white one gleamed, seemingly misplaced among its dark companions.

"Hey, fellows," Dave called from one side of the room, "look what we've got here. And remember, please be gentlemen."

We clustered around him to look at the low stalagmite his light pointed out at his feet.

"Boy! Even Mae West'd be proud to have a couple like that."

A perfect replica of a woman's breast stood there—softly contoured, proud and erect, in tan limestone that shone in our lights like softly polished marble. Many thick-based stalagmites stood about, several of which had joined their companion stalactite overhead to form sturdy columns. In our lights they glowed wetly, like terraced temples. One reminded us of a formation in Carlsbad Caverns which was called the Temple of the Sun, and we named this room after that one. But the name did not endure, even with us. Always thereafter we referred to this chamber as the Mae West Room.

A small passage led us from this area, sloping gradually downward, and we had to stoop most of the way. In about fifty yards we stepped into a low-ceilinged room in the center of which lay a beautiful little pool of water, the edges bordered with the flat rounded formations we called lily pads. Nowhere was the ceiling more than four feet above the floor. On one side stalactites hung so thickly to the floor they seemed a forest of brown tree trunks marching to the water's edge. The crystal-clear water reflected everything like a mirror.

After a long cold drink from the pool, which we called Lily Pad Spring, we continued down the corridor. It opened after about sixty yards into a larger passage about twenty feet wide and almost as deep, running at right angles to the corridor we had just quitted. We could see that the new passage ended a few feet to our left, so we turned up it to the right, soon finding ourselves at the foot of a steep slope where the passage took a sharp turn upward. We scrambled up the incline,

57

which was covered with loose dirt and fallen rocks, and found our way still sloping upward, but less steeply. The going was rough, cluttered with loose rocks and boulders that had fallen from above. The walls and ceiling were barren; nothing could be called beautiful here.

Suddenly Tommie stopped. "Whoa!" he exclaimed, pointing to a peculiarly curved stalactite which resembled a scimitar hanging from the ceiling near the right wall. "I've seen that before."

Beneath it at the base of the wall was the small hole we had squeezed into several hours previously. We had made a circle underground and were back in the main corridor.

"I can't believe it," Glenn said. "We've been down in a lower cave level."

"What I'd like to believe is that we can find our way out of here." Sonnie sounded skeptical. "I'm sure as hell turned around."

But we found our marking arrows scratched at intervals in the floor, and they led us finally to the soft glow of daylight coming through the entrance. We stepped outside to an atmosphere turned ruddy with blowing dust. Wind whipped the sotol and ocotillo in frenzied dance, and the sun seemed a burnished penny in the haze overhead. The encircling hills disappeared softly into the near distance.

"I think I'll go back into the cave," Tommie bitched.

"I'll say one thing," Glenn said. "We've got a lot of cave here to explore. But no more today—I've had enough for one time."

The lure of this new cave was irresistible, and one week later—on March 11, 1934—Dave and I were back. The other fellows could not come for one reason or another, and we had a tyro cave explorer with us, Ray Sims. We had just cleared the capstone arch over the entrance and were standing in the half darkness letting our eyes adjust, when I noticed a corridor branching immediately to the left. We had overlooked it on our first trip.

"Might as well see what's there," Dave said.

We soon found ourselves in a maze of barren tunnels, seldom more than ten feet high and not much more across, that wound about among themselves and branched off into other passages, to merge again at a further point, and then branch again. Often we would follow a curving passage to come unexpectedly upon our arrows in the floor. Then we knew we had just passed that way, and that what had seemed a branching corridor had only been the same one split into two corridors by a partition. There were no decorative formations, just the lifeless brown rock, and the passages seemed endless.

"Is there a better name for this than Endless Cave?" I asked. "If there ever was one without end, this is it."

Ray was not enjoying himself too much. "I thought you said this was a pretty cave. You call this pretty?" He pointed to the barren walls.

"O.K., Sam," Dave put in, "let's take him to something pretty."

We knew we were back quite close to the entrance corridor when a wide passage broke off at right angles to the left, and we decided to see if it might lead to something better. It did. After several yards we stepped into a long, flat, oval-shaped passage about ten feet high and twice as wide. A pool of water covered the center of this corridor; it was about ten feet wide and fifty feet long and in many places at least four feet deep. There were no lily pads here, nor stalactites reflected in the depths—only the pool and the brown walls. And something else. Two wooden barrels stood dejectedly at the water's edge, their seams open and the metal binding bands loose and askew. An old wooden packing crate slouched on end at a crazy angle. Several coils of copper tubing lay about. An old single burner kerosene stove tottered on three legs, rusted and disintegrating.

"So this is where old Crazy Ned used to make his bootleg!" Dave exclaimed. It was the remains of an old illegal corn whiskey distillery lying about, long since abandoned and forgotten.

"Don't you know this room smelled to high heaven when he got his mash cooking!" Ray mused.

We went back to the main portion of the cave and finally led Ray to the Grand Canyon by the Lily Pad Spring route. As we passed Taffy Hill, Ray brightened up, "Well, this is more like it."

As we walked by the Totem Pole, I noticed a rather large opening leading down right against the wall. I do not know how we had missed it on the previous trip. We eased ourselves down through the hole into a narrow passage that bent sharply to the right and opened within a few yards into the most beautiful room we had found in this cave. It was roughly fifty feet square. The ceiling, averaging a dozen

58

feet overhead, was lost in a maze of slender stalactites, snow-white to chocolate brown, in all lengths to as much as four feet. Most of them glistened with seeping water, and nearly every tip held a gleaming drop. Almost the entire floor surface either was or had been the basin of an underground lake. Lily-pad formations lined the walls where the waters had stood, and coral-like masses reared up like fountains to the old water line, as level on top as the surface of the water had been. All were a deep brown, rough and spongy looking. At almost every point that had remained above the old water level stalagmites had formed, from white to dark sienna. Many extended clear to the ceiling, like supporting columns—some were as big around as my leg. The walls were covered with rippling brown flowstone curtains, and long rows of sepia stalactites hung beneath every projecting ledge.

The floor of the right half of the room was about five feet lower than the rest, and a pool of clear water filled the area to a depth of four feet. Slender stalactites no larger around than a pencil hung by the hundreds over the water; many were four or five feet long, and we saw several we estimated to be ten feet long. Many had broken, probably from their own weight, and when we examined the broken pieces we found they were hollow, like soda straws. The long ones still hanging were so delicate that we could see them sway slightly just from the reverberations of our talking.

"We need some frosted cokes to make this the Soda Straw Room," Dave remarked. "I never expected to see anything like this."

"Well, Ray, is this pretty enough for you?" I asked.

"Prettier'n a picture, Robert . . . prettier'n a picture."

On our way out of the cave we found a small hole near First Spring that dropped into a small series of chambers. Their white flowstone walls were roughly decorated in many places with popcorn-like protuberances. Gleaming white clusters of stalactites hung from the ceilings, looking much less like stalactites than like hanging bunches of grapes.

Two weekends later found Tommie, Dave, Glenn, and me driving through a blinding snowstorm on our way to Endless Cave. We drew lots, and Glenn and Dave got stuck with the rumble seat out in the weather.

"Fine damned thing," Dave grouched. "Three weeks ago I choked back here in a sandstorm, and today I'm freezing to death in a blizzard."

On our way back to the virgin beauty of the Soda Straw Room, we decided to see what might lie concealed beyond the rat hole we had noticed on our other trip at the base of the drop just beyond First Spring. We immediately found ourselves in a maze of tunnels that branched in every direction, doubling about and circling back upon themselves. They were barren of limestone decorations, but many weird-looking erosion remnants of white gypsum stood about, like gnomes in our lights. No corridor was more than twelve feet high, and some were as wide as eighteen feet.

"This reminds me of the Badlands. There's certainly little good about it," Tommie complained. I could tell he was thinking of the beauties around the Grand Canyon area.

We decided to work out a few more of the passages, and it seemed that Tommie's wishes must have guided us, for within a short distance we came out of a small hole right at the Grand Canyon.

We emerged from the cave into a scene almost as unreal as those we had just left. The whole world lay motionless under a shroud of white. The snow had ceased, as had the wind, and every bush, every cactus, every shrub bowed beneath its burden of clinging white crystals. The sun glowed faintly in a lowering grey sky. Old Man Winter had had one last shout, and all the world was quiet.

On another trip several months later we worked our way along a tunnel that branched to the left at First Spring. It led into a series of chambers beautifully decorated with brown flowstone walls. Even the floors were covered with the material. Here, again, a bewildering maze of passages seemed to wander in all directions. We kept on working them out, and eventually came back into the main corridor deep within the cave. Endless Cave . . . how appropriately we had named it.

On still another day, Sonnie Hagler and I decided to take our girl friends through the cave. His current flame was Mary Nell Berger, and visiting in my home was Marge Dwyer, a college friend from Silex, Missouri. Far down a tiny corridor leading from the Soda Straw Room we clambered into a large chamber strikingly different from anything we had found. It apparently had once been almost filled with

Endless Cave 85

water, and must have remained that way for a long time, as the old water line showed clearly on the walls several feet above our heads near the ceiling. But the water was gone now. Strange, porous-looking stalagmites reared up from the floor to a height of four or more feet. They looked like giant elongated sponges or spires of jagged coral. Some were as big around as my body.

Snow-white stalactites hung from the ceiling, and where they had touched the old water level great round masses of the coral-like material had formed on the tips, making grotesque war clubs. Some of the knobs seemed much too large for the stalactites to support them. The room was rectangular, about fifty by a hundred feet. Everything below the old water line was a deep reddish brown, while everything above it was a gleaming white. The contrast was startling.

"I saw an underwater movie one time," Marge said, "and I remember it had some scenes in what they called the coral gardens. This looks just like it."

"Coral Room . . . " Mary Nell mused. "I like that."

I made several more trips into the cave, each one bringing pleasure and new discoveries. Then I went away to college and my cave exploring suffered, save for one or two trips during summer vacations. Following graduation, one of the first things I did upon returning home was to go by Tommie's house to get up a cave trip.

"Suits me. Where'll we go?"

"How about Endless Cave?" I asked.

"I don't know, Sam," he hedged. "I don't think you want to go there."

"Not go to Endless?" The very thought was preposterous.

"I kinda hate to tell you this. You know that guy that polishes and sells cave formations?"

"The one who offered me twenty-five dollars to take him to Endless Cave?"

"Yeah." Tommie's voice mirrored his disgust. "Well, someone sold out to him for five dollars and took him to the cave. And, Sam, he wrecked it. There's hardly a formation hanging or standing. He just stripped it clean. Anything that could be sledge-hammered down is gone. The only thing worth looking at is the Coral Room—somehow he missed it."

"Not the Soda Straw Room . . ."

"Here," he answered. "Let me show you something."

He reached up on a shelf and handed me an enlarged photograph. It took me a while to recognize the scene. Then I realized it was the Soda Straw Room. But nothing I remembered was there—just bare walls and ceiling and a few remnants of the coral fountains and broken edges of the lily pads. Gone were the soda straws and the gleaming white stalactites and the rows of chocolate brown draperies that had festooned the walls.

I realized my hands were shaking. I felt like a little boy holding his favorite toy that had been smashed by the bully next door.

The only difference was that the little boy could cry.

I never went back to Endless Cave.

8
Cottonwood Cave

OUR TRIP TO Cottonwood Cave came about almost accidentally. About a month previously, Dave Wilson, Ted Fullerton, Julian Shattuck, and I had visited Black Cave far up in the Guadalupes, and on the way back we had discovered a new cave we named Hell Below. We had gone as far into it as was possible without rope, and left determined to come back at a later date with sufficient rope to explore its depths. So on September 10, 1933, the four of us and Tommie Futch, who had joined our party in Carlsbad, left Julian's ranch on horseback, headed for Hell Below.

Our way led gradually upward for several miles, then dropped off into Dark Canyon, which we followed for another mile; then we began our ascent of the steep climb in the vicinity of the Dark Canyon fire lookout tower. We pulled our horses up for a breather; the tower stood only a short distance to our right, up on the high ridge.

"While we're this close," Julian said, "let's ride on over to the lookout tower. I want to show you the view off into Black Canyon."

We stopped at the lookout tower and let the horses have a drink, then rode out to the edge of the ridge. The view off into Black Canyon almost took away our breath. Great jagged gashes dropped off beneath us, matching similar ones plunging down from the far side of the canyon, tinged blue in a morning haze. Massive pillars of rock thrust up from the dropping ridges like towers in the sky, and vast precipitous bluffs plunged toward the canyon's floor. The canyon made a bend far below and to our left and disappeared behind a shoulder of its own north slope.

"Hey," Julian said, "Cottonwood Cave is just over there. Why don't we take a look at it before we go on?"

"You sure we have time?" Ted asked. "It may take a while in Hell Below."

"Plenty of time. Cottonwood's just one big hole in the ground with some really great formations. You'll get a kick out of seeing them."

So we rode the short distance over to Cottonwood Cave and tied our horses in a grassy glade above the entrance. The walk downslope to the opening was only about a hundred yards. A black hole, some twenty-five feet across and a dozen feet high, gaped in the side of the slope. The steep drop-off into the mountain seemed almost vertical as we gathered about, checking our lights. A small oak stood squarely in the entrance.

"I know it's not a cottonwood," Julian replied to my question. "There are no cottonwoods up here. I never did hear how the cave got its name. Any damned fool would know that is an oak."

It seemed to me the cave floor dropped a hundred feet in about the same distance, and we found ourselves in a great auditorium that was faintly lighted by the daylight coming in from the entrance far above. Our voices echoed about the vast chamber, seeming to bounce back and forth from wall to wall. Great white stalagmites stood about, like huge monoliths in the dim light. Directly to our left a tall, graceful one tapered to a sharp point twenty feet above. A shorter one, bulky and knobby, about a dozen feet overall, with a flat top, stood squarely in our trail, while slightly behind it a shorter, stockier stalagmite, seemingly its companion, helped guard the passage.

66

"They look just like Mutt 'n Jeff," Dave pointed out.

Behind and to the right of Mutt and Jeff towered a real dandy. Terrace and fluted side, terrace and fluted side it soared up into the darkness. We could see that the top was almost flat, and from one edge of it dripping water had built up an extra two feet of stalagmite, like a pointing finger. Tommie thought we should call it the Finger of Doom. Measurements of this one revealed a base thickness of six feet, while the tip of the pointing finger seemed just under thirty feet overhead.

Forty feet beyond, just visible in the fading light, loomed another big one. It must have been ten feet through at the base, rising with almost parallel sides for about twenty feet, then tapering sharply to a point thirty feet above the floor. A stalactite had grown down from the ceiling and firmly attached itself to the tip of the big formation below. Behind this one and to the right stood another, not so graceful but fully as large, and farther back in the gloom loomed another almost as massive.

In places the walls bore cascades of brown flowstone. The floor was level and smooth. What few stalactites had formed on the ceiling were small and insignificant when compared with the giants standing beneath them.

"This is truly a Hall of the Giants," Ted remarked, trying to penetrate the shadows overhead with his feeble flashlight.

The width of the entrance auditorium seemed at least fifty feet, and it continued into the mountain, enlarging gradually as we walked along. The ceiling seventy feet overhead became lost in darkness as we left the daylight from the entrance. The great corridor sloped gradually downward for about twenty minutes; then we began a gradual ascent up a boulder-strewn slope. There were practically no decorative formations to be seen, just the vast hall, rock-filled and rugged.

After we had climbed gradually for sixty or seventy feet, the corridor belled out into a roughly circular room, with the floor sloping more steeply up toward the center. Our lights, like fireflies in an arena, illuminated little, but even in their feeble glow we could not have failed to see that great shape looming before us.

"I don't believe it," Tommie whispered.

Towering before us, up and up until its top disappeared in the shadows beyond our lights, stood the largest stalagmite we had ever seen. Its light-tan massiveness so dwarfed us we felt like pigmies gathered around its base. It dominated the center of the chamber, rising up in alternating sections of smooth terraces and graceful fluted draperies. Astonishingly symmetrical for something so massive, it hardly varied in diameter from bottom to just below the top, where it tapered abruptly to a rounded point still several feet beneath the ceiling.

"And I thought the Giant Dome in Carlsbad Caverns was big," Ted marveled. "How high do you think that thing is?"

We set about trying to measure it. We always carried a reel of twine to mark our way out of unfamiliar corridors, and we stretched a length of this entirely around the base of the great stalagmite. Measuring this by the old tip-of-nose to arm-length method, we arrived at a circumference of roughly seventy feet, or a diameter of approximately twenty-four feet. We could estimate its height only by comparison with six-foot Dave standing at its base, and we came up with somewhere around seventy-five feet.

"Sure as hell beats anything in the Caverns by a lot of rock," Dave commented. "If this isn't the world's biggest stalagmite, it sure will do until a bigger one comes along."

"We've got to take a picture," Ted burst out. "Can you get it?"

"If I used all the powder I've got," I said, "I don't think I could get it, and I want to save some for Hell Below."

But I decided to try, and poured out enough powder to have sufficed for three or four photographs under normal conditions. I could not back up far enough to get a level shot at the thing, so had to compromise by tilting my camera up at too sharp an angle.

"This will wreck the perspective," I wailed. It did. And the amount of powder was insufficient, even though the great chamber boomed when the charge went off. The top of the goliath all but disappeared, faint and underexposed, in my finished print. But even then, the massive proportions of the great stalagmite were apparent, in comparison with the figures of Dave and Ted sitting at its base.

Just beyond, the great corridor pitched off at about a 50-degree angle for more than a hundred yards. This slope was covered with a deep deposit of pure sand.

"Whee!" Julian yelled. "This is what I've been waiting for." He made a run and a great leap out onto the sand toboggan, then scooted for many yards, hands and legs in air, and sand flying.

"Beat you to the bottom!" Tommie shouted and started a dash down the slope. Ted and Dave and I followed, big steps getting greater as we picked up speed. Ted's head got faster than his feet, and even gargantuan steps could not save him; he plowed face first into the soft sand and slid to an ungraceful halt, coming up spitting dirt and expletives.

And so we played there on that great sandhill deep in the heart of the mountain, like children romping among sand dunes at the beach, with only our flashlights to light our way as we tore at breakneck speed down the slope into darkness. The great corridor echoed with our shouts and laughter. I guess, for a few moments at least, we *were* children again.

The corridor came to an end just a short distance beyond the base of the sand slide, and that was all we found of Cottonwood Cave. It remained for more careful explorers, determined to push the cave farther, to find, many years later, a vast and greater cavern on a lower level than the one we explored.

POSTSCRIPT

I returned to Cottonwood Cave on April 1, 1961, with the express purpose of seeing the newly discovered area on the lower level. The great auditorium with the massive stalagmites standing about were much as I remembered, and our goliath still stood, gigantic and unchanged, like a sentinel guarding the passage. The vast slope of sand still pitched off into darkness, but I bowed to the years and made no attempt to race someone to the bottom. Once I thought I caught a distant echo of laughter from a remote past, but it must have been only voices from some of our party ahead.

The trip into the new area was magnificent; some of the wonders hidden there staggered the imagination.

But that is a different story from another time.

9
Goat Cave, Grey's Cave, Gunsight Cave, and Gyp Cave

Goat Cave

GOAT CAVE HAS THE KIND OF ENTRANCE that promises great things when seen across the canyon, and yields precious little when you finally clamber to it. I had camped in the south fork of Slaughter Canyon during the 1933 deer-hunting season, and late one afternoon as I was headed for camp I attempted a shortcut across an intervening ridge. As I topped out on the ridge, a large black hole loomed up on the opposite slope. Even from my distance of more than a mile I could tell the entrance was huge, but my binoculars revealed no details in the darkness beyond the opening.

"I'm going to mark this one well," I thought, "for we've got a cave to explore here."

It was not until a cold January 7th of the next year that Dave Wilson, Tommie Futch, Glenn Hamblen, Ted Fullerton, and I finally got around to looking for the cave. After parking the car at the extreme end of the road, we started walking up the canyon. Occasional pools of water left from melting snow that had fallen several days previously were iced over. The shallower ones were solid. We had some fun at impromptu ice skating on some of the larger ones. Of course, ice skates would have helped. With a flourish Dave started a grand slide across one and ended up on his backside at the far end of the pond, his glasses, cap, and pack flying in all directions.

"The prima donna of the ice opera," Tommie taunted.

Ted had bet that I would not be able to find the cave, but for once I won. I turned up the side canyon to the right, and soon the black hole yawned in the slope above us. The climb was not extremely difficult, and as we drew nearer we began to realize how tremendous the opening was. When we finally stood on the level bench stretching back into darkness, the great arch soared fifty feet overhead and spanned a good 150 feet. Several trees with six-inch trunks grew at the entrance, and their topmost branches lacked several feet of touching the arch above.

"If the size of this opening means anything," Glenn remarked, "we should have a lot of cave to explore here."

The floor of the entrance chamber extended back as level as a ballroom. A wealth of daylight flooded the great room, and we could see the curve of the ceiling stretching back for fifty feet to meet the floor. Algae had colored the walls and ceiling a delicate grayish blue near the entrance. This shaded to a baby pink toward the back. At the left center a closely grouped cluster of brown stalagmites about six feet high broke the monotony of the flat floor. Stalactites up to six feet long dotted the ceiling here and there; some of them were massive and blunt, lacking the taper found in most.

"If that dark tunnel back at the right doesn't lead somewhere," Dave suggested, "this cave is going to be just one big hole in the wall."

A goat herder had built a sheep-wire fence across the entrance. He could not have been too ambitious; the few posts he had set in the rocks leaned at all angles, and he had utilized the trunks of the growing trees to serve as others.

"Probably used this as a shelter to pen his flock," Ted mused.

"The shape it's in now, it wouldn't hold many goats."

We scrambled over the dilapidated barrier and walked into the chamber. The completely smooth and level floor amazed us. Our voices echoed back and forth from wall to wall. The black tunnel at the back that Dave had indicated seemed a promising lead, breaking the symmetrical downsweep of the ceiling with a black opening twelve feet in diameter. The floor of the tunnel sloped sharply upward, and Glenn was the first to try it, scrambling up out of sight.

"It goes on," he called down. We started up the slope after him.

"It stops," he said, disgustedly, about the time we came in view of his light.

The tunnel ended in a roughly circular room about twelve feet across, with the ceiling disappearing into a dome forty feet overhead.

A few taffy-colored stalactites hovered up there in the darkness. The walls were one solid mass of brown flowstone; when water had trickled over them they must have been beautiful, but now they were dry and dull. A column as large around as a man and twelve feet high grew from a mound at the back of the room and joined the lip of a protruding ledge that encircled the dome. We turned out our lights at the back of the room, and as our eyes became accustomed to the darkness, phantom fingers of daylight crept up the pitching floor of the tunnel to bathe the chamber finally in a soft, blue glow that faded gradually into the pitch black of the crypt overhead.

"Sure would like to get up there," Ted remarked, flashing his light around in the dome. "I'll bet there's a hole up there somewhere that leads to more cave." But the walls were smooth and the handholds precarious, so none of us dared attempt it.

As we scrambled through the fence and left the cave, we stopped and looked back. The awesome span of the opening still fascinated us, although we knew the reaches behind held little of interest.

"I guess we ought to think up some appropriate name," I said. "Something like Blue Arch Cave or Lost Dome Cave. But for some reason, all I can think of is Goat Cave."

"Sounds good enough to me," Tommie added. "The goats had it first, and they'll probably be back after we're long gone. Goat Cave it is."

Grey's Cave

I suppose there are few areas anywhere on earth boasting of caverns that do not have at least one lost cave or mystery cave. The Guadalupes have their Grey's Cave. I know it is there, or at least I know it existed once, for I have four photographs to prove it. When pioneer Carlsbad resident Elliot Hendricks died, he left among his possessions a box of old photographs, mostly portraits of relatives and friends. But in the collection were four pictures, 4 × 5 inches in size, mounted simply on plain cardboard, with no identifying inscription on the front. Two of them pictured a group of persons in rocky terrain, and on the back of the cardboard mount was written simply "On way to Grey's Cave, Carlsbad, New Mexico." The other two were apparently made with flash powder inside a cave, and pictured the same group of people gathered about amid white stalagmites and stalactites. The backs of these bore the inscription "Grey's Cave, Carlsbad, New Mexico."

The mode of dress of the people involved would date the pictures at about the turn of the century. The ladies were decked out in long skirts that swept the ground, puffed-sleeve blouses, and neck ribbons. One even sported fancy gloves and a perky straw hat with veil. A little boy about ten years old wore suspenders and baggy pants, a fantastically flowing box tie, and a wilted fedora. The men were elegant in jacket and vest, stiff collars and bow ties, and English caps. One carried what seems to be a 30-30 carbine, and the other toted a double-barreled shotgun. If they expected bears in the cave, they were prepared.

73,74

I have asked all the old-timers I could contact, and none of them ever heard of Grey's Cave. One lead I thought would surely produce was to a pioneer ranching family named Grey who settled near the mouth of Big Canyon across the line in Texas, but the daughter who lives in Carlsbad had no ideas at all about it.

The countryside in the photographs does not suggest the high mountains. It more nearly resembles the cactus-covered foothills found in the area around McKittrick and Endless Caves. But none of the cave formations pictured in the two flash photographs resemble in any way those seen in early photographs of McKittrick Cave, which was the popular picnic spot of the era.

I guess Grey's Cave will have to remain a mystery cave until that time in the future when some lucky caver stumbles across it. I wonder if some of those old white tallow candles the early explorers carried will still be stuck on the stalagmites where they left them.

Gunsight Cave

We made a try for Gunsight Cave the morning that Tommie Futch rescued Dave Wilson, the two Canadians, and me from Spider Cave. We got out of that mess about daybreak and got back to town just in time to start for Gunsight Canyon, where Mrs. M. E. Riley, an old-

75

time resident, was going to lead us to a "big cave." But the walk from where we had to park the cars near the entrance of the canyon to the cave far back within was just too far and too rough, and we were just too tired from the experiences of the night before in Spider. Mrs. Riley assured us that all we had to do was continue up the canyon and we could not miss the big hole away up on the right slope.

So the next weekend, April 22, 1934, found Dave Wilson, the Canadians Fred Burgess and Cal Livingston, and me trudging again the rugged floor of Gunsight Canyon. As we wound our way up the rocky washes, the walls grew higher and steeper about us. Undergrowth choked the dry stream bed, and in many places we seemed to fight our way through. Finally the canyon made a bend, and we saw the entrance to a cave, away up high near the rim. It was definitely a big cave, as Mrs. Riley had promised.

76

"Thank heavens we didn't try to make it the day after Spider," Fred said, wiping away the sweat. "I don't know if I can even make it today, and I had a good rest last night."

The climb up out of the canyon was a grueling struggle. We soon left the undergrowth and its clinging arms, but the barren ledges and loose-gravel slopes were no improvement. Thorny cactus lurked everywhere, waiting for a tussle. The steepness of the slope ahead often hid the great entrance above; then we would catch glimpses of it, growing ever larger as we climbed toward it. When we finally rounded the last shoulder and the great hole yawned before us, we could not believe its size.

"Damned if they couldn't fly the biggest dirigible in there," Dave exclaimed, "and have room left for a couple of airplanes alongside!"

"Never did I think I would see a hole like that," Cal said.

"It must be a hundred feet from top to bottom," Fred put in.

"And half as wide across."

The cave dropped off immediately into the depths. The slope down was steep, rough, and boulder-strewn. We made our way past a solid pinnacle of rocks twelve feet high that stood squarely in the center of the opening. A pair of oak trees thirty feet tall shaded the pinnacle but were dwarfed to shrubs by the towering arch overhead. After about seventy feet of slipping and sliding, we reached the floor of the cave, which leveled off into three branching corridors whose ceilings towered high above us. No flashlights were necessary here, for daylight poured in through the vast entrance up above.

"Which way shall we go?" Dave asked. "It looks like we've got more cave here than we can explore in a week."

But his optimism was short-lived. One after another the corridors, starting in such grand proportions, ended in blank walls. None of them went beyond the reaches of daylight. We scrambled up the left one first, finally coming to a halt where it ended high up near the great dome of the ceiling. On a deep ledge here stood a massive stalagmite, rising in a couple of terraces to a sharp peak eighteen feet above. It stood in dark majesty, almost black in color, and we called it the Black Pagoda.

This corridor showed signs of great stress. An apparent settling of the big arch had forced columns out of line, and in many places had snapped them in two. Some leaned out crazily from the wall behind, and many had fallen in splintered pieces to the floor. Huge boulders littered the way, and we could see the gashes overhead from which they had crashed.

"Been some trouble here," Dave worried. "Things sure started cracking up."

A great domed stalagmite with fluted sides stood guard in front of the right corridor, but it really should not have bothered, for there was nothing there to see. It would have served a better purpose in front of the middle corridor, for here the floor swept up at such a sharp angle we could hardly climb it, ending in a great domed room thirty feet across with the ceiling dim in the gloom overhead. Down the back wall poured a frozen cascade of brown flowstone, fluted and terraced, apparently emerging from a horizontal cleft high up near the ceiling. At the right of the room a great column, actually a colossus twelve feet through at the base, reared up in scarcely diminishing diameter, seeming to support the ceiling far above. It, too, was fluted and beautiful, of a dark brown color, except in perpendicular sections where a topping as white as whipped cream seemed to have flowed down from above.

The cleft above the flowstone waterfall offered possibilities of exploration, but we lacked ropes or any other aid for careful climbing, and one look was proof that any climbing here would have to be careful.

"Some day I'm coming back," Dave said, "and I'm going to see if there isn't a second floor to this cave up there above that waterfall."

As we started up the steep slope out of the entrance, the pinnacle of rocks up above stood out in stark silhouette against the skyline, like the blunt front sight of an old carbine nestling in the notch of the great entrance.

"Look there," I exclaimed, "just like the sights on Dad's old 30-30! A Gunsight Cave in Gunsight Canyon. What better name could we find?"

I went back to Gunsight Cave on March 19, 1939, with Seth McCollum and Bill Burnet. Bill was an amateur archeologist and pale-ontologist of no mean ability, and he had heard that someone had found a deposit of bones far back in the cave. When one mentioned old bones to Bill, a terrier could not have been more interested, so he talked Seth and me into taking him.

On this trip, we did not attempt the terrific climb up from the mouth of the canyon; instead, we drove up on top of the Guadalupes, as far as we could beyond the Dark Canyon fire lookout, then walked down the ridge overlooking Gunsight Canyon on the right. It was a much easier trip, and the great entrance was no less impressive when we came out on the overhanging ledge and it dropped away beneath our feet.

"Overwhelming!" Bill exclaimed. "Absolutely overwhelming."

He found his bones, deep down and back against the far wall. A great boulder had fallen from the ceiling and partly overlay the site, but Bill managed to dig out a respectable pile that Fido would have been proud of. Bill was certainly proud of them. He spouted out a Latin name or two that meant nothing to us; I learned later that one of them meant "three-toed sloth."

Seth spent the day prowling every available section of the cave, except the horizontal clefts high above the rippling flowstone.

"Too steep and too slick for me," he explained, when I asked him why he did not try it.

I spent my time taking pictures. I was trying out my new Eastman Recomar camera, and like a child with a new toy, I put it through its paces, seeing what I could get in the available light pouring down through the vast entrance.

For each of us, it was a day well spent.

Gyp Cave

Our expedition to Gyp Cave was one of those things cavers sometimes do to while away an otherwise wasted Sunday afternoon. This was February 25, 1934, and what had promised to be a perfectly lousy day turned out balmy and warm by mid-afternoon.

"We oughta be in a cave somewhere," Dave complained.

"I've got just the one," I told him. "There are some sinkholes a few miles out of town, off the road to Spencer Dam. Some guy told me about them the other day; he claims he found them while rabbit hunting, and he doesn't think anyone has ever gone down in them."

We picked up Tommie Futch and a length or two of rope and headed into the low, cactus-studded range of hills that rise up west of Carlsbad in gentle slopes, then fall off again into the valley beyond Rocky Arroyo. I had only a vague idea of where the caves were, but we parked the car about where I thought the man had told me to, and started walking up the slope to the north. With the luck that sometimes blesses cave explorers, we came right up on three pits that dropped almost straight down through the rocks. It took only a moment's inspection to show that two of them did not amount to anything, but the third one presented a different story.

This one was like a fissure in the rocks, which were loose and crumbly around the opening. About five feet high and ten feet wide, it dropped straight for about fifteen feet, then made an abrupt bend back under the entrance and disappeared into blackness. We pitched some rocks off, and they bounded down out of sight, sending up rumbling echoes as they clattered into the darkness beneath us.

"Sounds like there may be quite a hole here," Tommie said. "String the rope, Dave, and let's go have a look."

I was not too fond of the crumbly appearance of the rocks around the opening, so found a good excuse to remain topside. Someone ought to stay outside to man the ropes and be there in case help was needed, I explained.

"Come on, Nymeyer," Dave scoffed. "Just admit you're scared to go down."

"Not scared, just cautious."

They knotted the two ropes together, and Tommie slid down, feet propped against the wall, to the first landing where the cave bent back beneath us.

83

"What do you see?" Dave called down.

"Lots of black."

Dave slid down to join him, and Tommie disappeared into the darkness beneath. In a few moments Dave took a new hold on the rope and leaned over to peer after Tom, and the rope dislodged a rock from the crumbling wall. I saw it tumble past Dave and bounce into the cavity beneath. Echoes came pouring out of the hole, and some of them seemed to be from a human source.

84

"Did Tommie get hit?" I asked.

"I think it missed him."

"What's all that noise about?"

"He's just talking about it."

Before Dave could go down after him, Tommie's head poked up from beneath the overhang.

"What did you find?"

"Well," Tommie drawled slowly, "I found a room about the size of Grandma's kitchen, a bunch of rusty cans, parts of an old cast-iron stove, the banged-up wheel from a flivver, and the skeleton of a dog."

As we pulled the rope from the cave and coiled it for carrying, Dave asked, "What shall we name this one?"

"How about Gyp Cave?" I asked.

"Gyp is a kind of rock. Would it be appropriate here? I guess some of this rock is gypsum."

"It's also what you get in a crooked crap game."

"Yeah, a lot of nothing."

"O.K. . . . Gyp Cave."

Goat, Grey's, Gunsight, and Gyp Caves

10
Chimney Cave, Dry Cave, and Hamblen's Cave

"THAT'S GOING TO BE ANOTHER of my famous tight squeezes," I remarked, taking a dubious look at the narrow cleft dropping straight down into darkness. The opening looked as though some giant of old had taken a mighty axe and split the huge white rock with one blow.

"It will be tight," Sonnie replied, "but I think you can make it."

Today, October 9, 1938, was our second attempt to find a cave someone had told Sonnie Kindel about. His first directions had been vague: "Right on top of the ridge a little ways east of Kindel Cave. You can't miss it."

We *had* missed it the weekend before, and had paced off several weary miles over the rough rocks without success. During the week, Sonnie had talked with his informant again and received more precise directions. Now we were there, having found the opening easily. We had missed it previously by a scant hundred yards.

"That must be the old ladder Jim White put in there long ago," I said. "Do you suppose it's still safe?"

Made of two strands of galvanized wire interlaced between twelve-inch lengths of one-inch pipe, it looked safe enough. The end was securely anchored around a boulder a few feet from the hole. Sonnie leaned over and shook the ladder, peering down into the crack and flashing his light about.

"Can't see a damned thing down there," he muttered, "but the ladder seems secure enough. Old Jim made 'em strong when he made 'em."

"Just the same, I'm going to use our rope alongside, just to be safe." We had brought a hundred feet of rancher's stake rope, and we knotted it around the boulder with the ladder and dropped the coils down into the cleft. We could hear quite a bit of the line's end strike the bottom of the cave.

"How deep do you figure she is?"

"Oh, about fifty or sixty feet I imagine. There's several feet of rope on the rocks."

"You'd better take her first, Sam," Sonnie suggested, "so in case you get stuck here in the entrance I can help you out."

"Or so in case I get part way down and the ladder breaks, it won't be you that hits the bottom!"

I looped the safety rope under one arm, across my back, and under the other, and eased into the crack, clinging to the ladder. It was a tight squeeze, but after a few strategic wriggles and some violent exhaling of breath I felt my feet and rear end clear the embracing rocks and break into space. The thickness of the rock crust was only about six feet.

As soon as I cleared the opening and swung free, I started having trouble with the ladder. Its bottom end was not attached to keep it rigid, so my feet, instead of staying beneath me where they belonged, stuck out at right angles about on a level with my belly. The cursed thing swayed in every direction, and the "safety" rope almost hanged me before I had done with the damned thing and cast it off.

"Why don't you smarten up and quit getting into these damned messes?" I fussed, talking to myself.

Somehow I made it to the bottom, and a great hall of blackness engulfed me. The crack above made a narrow gash of light in the void, and I heard a flutter of bat wings and saw the creatures in darting silhouette far up against the glow of daylight.

"How'd she go?" Sonnie called down. His voice echoed hollowly around the auditorium.

"Like a descent into hell, but come on down."

In a few minutes he broke clear of the gash, and almost immediately his feet shot out horizontally; his cap fell off and came spinning down.

"Jesus damn!" I heard him sputter. A flashlight broke free of his pocket and flew into pieces on the rocks beneath. Then the ladder started a slow spin, and the dangling rope wrapped around his ankles.

He twisted violently to stop the spin; the whirling suddenly reversed, and one foot slipped from the pipe rung and popped through the ladder.

"For Christ's sake, Nymeyer, grab the damned ladder, or do something!"

By then I was laughing so hard I could barely make my way over to assist, keeping a wary eye peeled upward for anything else that might fly loose and come hurtling down. But I managed to grab the lurching ladder, pull it taut, and stand on a rung. Sonnie worked himself out of the tangle and came on down.

"Boy, would I like to have a movie of that," I laughed. "The daring young man on the flying trapeze."

"Don't expect any sympathy the next time I catch you in a tight spot."

We started an inspection of the big chamber. It seemed almost circular in the dim rays of our flashlights, and vaulted out in a flat-topped dome. Almost in its center gleamed the gash of daylight. The floor was a great breakdown of rubble that had scaled away from above through the centuries; several well-developed cracks in the dome suggested that one of these days the final breakthrough would occur in the thin shell overhead, and Chimney Cave would become a great pit.

There just was not enough roof overhead to trap water and develop seeps for the growth of stalactites. This was one of the most barren of caves. We felt as though we were walking inside a vast bubble in the earth.

"Hey, here's something pretty," Sonnie called from away over at one side of the room.

Close against the wall a small tunnel led off for a few feet, and inside were some of the prettiest elephant-ear draperies we had ever seen. They were about two feet long and almost touched the gleaming flowstone floor, wet and glistening from the only seeping water in the cave. The curtains fell in graceful pleats and glowed a deep orange color in our lights; when a light was held behind them, they almost seemed to blaze.

"Sure glad to find something in this cave to photograph," I growled.

As we gathered back at the foot of the ladder, I suggested to

86

Sonnie, "Why don't you climb about half way up and let me get a picture of as much of the whole room as I can?"

"If you'll stand on the end, and no funny business."

"No use in my doing that. It seems to me this boulder here was just made for anchoring the ladder. I really think Jim had it tied here at one time."

We bound the ladder securely to the rock, and while Sonnie started his climb, I poured out a charge of flash powder almost beneath the rope against a pile of boulders, then walked back to the end of the room and set up my camera and opened the shutter. By the time I got back to the powder, Sonnie was halfway up and ready to pose. The flash and boom of the exploding powder started a great circling of bats that roosted above a projecting ledge just under the dome. Their staccato squeaks filled the room with uneasy sound. Sonnie came back down the ladder.

"Don't know why in the hell I didn't load up my stuff and take it with me when I went up; then I could have gone on out," he complained.

"I know," I replied. "You just wanted me to go first, so if I got stuck in the crack up there, you could stand down here and laugh."

I loaded up and started the climb out, but had gone up only a couple of rungs when a strange roar permeated the room. It came from no discernible place and was identifiable from no certain source; it was just a rumble that drifted about and softly died.

"What the hell was that?" Sonnie asked.

"I don't know. Did you see a flash of light, or did I imagine it?"

"I'm not sure. I thought I saw something. I know I heard that noise."

We listened, but the silence was broken only by the diminishing chatter of the bats settling back in their rookery. Then it came again, and this time we both saw it.

"That was certainly a flare of light up at the entrance." I had hardly spoken the words when the rumbling roar came again, louder this time, pulsating about the great auditorium. Then Sonnie slapped his leg.

"I'll be damned!" he exclaimed. "You know what that is? That's thunder."

"Thunder! There was hardly a cloud in the sky when we came in."

"Just the same, I'll bet when you get to the top you'll find a storm outside."

I continued on up the ladder. I carried my favorite caving light, an electric lantern using a big, square battery that weighed close to a pound. Feeling I would need both hands for the climb out, I had run my belt through the light's handle and let it hang behind. Anchoring the ladder at the bottom made the trip up much easier, and I soon reached the lower edge of the crevice. Then I had the worrisome problem of securing fingerholds around the rungs that were pulled tight against the face of the rock. While I was struggling with that, I heard a plop below and an exclamation from Sonnie, who was standing at the foot of the ladder helping to steady it.

"Jeez, if that had hit me, I'd be a goner!"

The bottom of my lantern had come loose, and the heavy battery had hurtled to the floor of the cave, missing Sonnie by inches. I squirmed on out, discovering in the process a second reason why carrying the lantern on my belt had been a bad idea; the darned thing caught in the narrow crack, and I almost lost my belt and pants getting loose.

As I poked my head up through the crack, I stared a bit of weather right in the face. A great, towering thunderhead had built up and piled in out of the east. It loomed almost straight overhead. Occasional drops of rain spattered the rocks. Just as I stepped free and hitched my drooping trousers back into position, a bolt of lightning out of the cloud crackled straight over me, and crashed into the ridge just to the west. The thunder boomed so rapidly on its heels that I dodged both at the same time and felt inclined to dive back into the hole.

Sonnie joined me just as the wind whooshed in, with a heavier pattern of raindrops riding the drafts.

"We're going to get our fannies wet," Sonnie predicted.

But the wind whooped on up the canyon, and the thunderhead marched in majestic splendor behind, stingily hoarding its moisture for the hills beyond. Brilliant blue sky swept up from the east to the zenith, and in minutes the sun peeked over the silver edge of the cloud diminishing to the west.

Chimney, Dry, and Hamblen's Caves   109

"Seems a shame to leave that good ladder hanging there," Sonnie commented. "Wonder why Jim went to all that work, found nothing down there, and went off and left it?"

"Probably the same reason we're going to. Unless you want to go back down and untie the bottom end."

By the time we reached the car parked about a mile away, the thunderhead was a horizontal bank of brooding white nestled up against the distant mountains, with the slopes beneath stretching soft, purple fingers toward the approaching sunshine. Lightning still bared its fangs in the cloud canyons, and thunder, like a great cat, prowled the high crags; but the storm, for us, had been merely a tumult and a shouting, signifying nothing.

Dry Cave

"Instead of going to Endless tomorrow," Sonnie Kindel said to Dave Wilson and me as we sipped coffee late one afternoon in a local beanery, "why don't we go look for that cave the hunter told us was just over the ridge beyond McKittrick Cave? If we don't find it, or if it doesn't amount to anything, we can go on over to Endless."

"Sounds like a good idea. Maybe we ought to give Endless a rest, anyway."

So the next morning, April 8, 1934, found the three of us spreading out along the ridge above and west of McKittrick Cave. Again luck was with us, for within thirty minutes I heard a shout from Sonnie a couple of hundred yards to my left, "Here it is."

Dave and I joined him. He stood before quite a gash in the hill. It spanned about twenty feet, with a perpendicular height of about eight feet from the entrance floor to the flat rock ledge that formed the roof. There seemed to be no more than six feet of rock above the opening. It yawned, black and rather foreboding, almost hidden by a heavy growth of brush.

The walk into the entrance was not difficult, with the corridor pinching down somewhat and dropping gradually into the hill. Within a few yards the passage enlarged into a room about twenty by sixty feet with ceiling around twelve feet high.

"An Endless Cave it isn't," Dave mused dryly, flashing his light around the barren rock walls and ceiling.

88

The floor continued sloping down, and we entered a corridor that rapidly narrowed to a tunnel that presented pretty tight going at times, and wound about in a couple or three switchback turns. Then it enlarged slightly and straightened out, and in fifty feet we stopped, seemingly on the edge of an abyss.

"This may be as far as we go," Dave, who was leading, commented.

As we shone our lights about, we discovered we were high on what might be termed a balcony, overlooking an almost circular room, with a domed ceiling rising above us, and the floor twenty feet below. The drop was perpendicular.

"Sure glad I brought my rope," Dave remarked, as he began uncoiling it from around his shoulders.

"You sure you're going down there?" I asked, leaning over and shining my light around. "That's a straight drop, and it seems to me the cave goes right back beneath us."

"I thought you came to see a cave," Sonnie taunted.

"What's the matter, Sam? You still remembering Hell Below?"

"I remember it well enough to not go hell-bent into something without checking first," I replied testily. Actually, Hell Below was still a vivid memory, and I could not work up a lot of enthusiasm for anything having to do with climbing a rope.

"I can tie her right around this projection here," Dave said, ignoring my remark and shaking out the rope. He made it fast and tossed the coils over the edge. He peered over with his flashlight. "Lots of rope on the floor."

"I tell you, fellows," I said, "You two go on down and I'll stay up here, just in case. If anything should happen, you might be glad to have someone up on top to get help."

"I still think he's scared," Dave said to Sonnie, "but it is a good idea for someone to stay up here. If we find anything worthwhile, Sam, we'll come back and get you."

"That's good enough for me," I replied, stretching out on the floor for a rest.

They slid to the floor below, and I could see their lights flickering about before they blinked out. It seemed they went off in a direction almost directly below me. I turned off my light to save the batteries, and the unbroken quiet of the cave settled about me. There was no dripping of water here to break the soundless monotony, nothing but the utter blackness and the boundless hush. I caught myself straining for just one sound.

I guess I dropped off to sleep, for suddenly I found myself sitting bolt upright. Then I realized there was a clatter and the sound of voices below, and a flickering of light. It was Dave and Sonnie.

"What did you find?" I called down.

"Nothing but dry cave and small rooms," Sonnie answered.

"This is about as dull a bastard as you can imagine," Dave grumped. "Not one stalactite to break the monotony."

Sonnie came monkey-fashion up the rope, and I grabbed his outstretched hand and yanked him up beside me.

"We did find a bunch of bones," he said. "There's some right down beneath us. I think it's the skeleton of a man."

"A man you think! Didn't you see his skull?"

"No skull that we could find. Just a bunch of half-buried bones."

"Probably a coyote that wandered in and fell over the edge."

"Too big for that. Could be a man."

Dave puffed up over the edge, and we helped him alongside. "If it's a man, I'll bet someone did him in one dark night, and brought the body here and pitched it over the edge to hide it. Pretty damned good hiding place."

"You think you found all the cave?" I asked.

"Well, there might be some tunnels that go somewhere. We saw one or two branching off, but they didn't look very promising. I don't think this is much more than a barren, dry rat hole." Dave shrugged the cave off as worthless and started coiling his rope.

As we walked out of the entrance and turned to look back at the black hole that had held such promising possibilities, I asked, "What are we going to name it?"

"How about the Dry Sonofabitch?" Dave grumbled.

We settled for just Dry Cave.

Off to the right about fifty steps we found another small hole, not much more than a crawl-in tunnel. Dave bellied down in front of it and shone his light in.

"Might as well see what's here," he commented, and disappeared. In a few minutes he came back out, dirty and sweating.

"Big cave?" Sonnie asked.

"Big nothing," he retorted. "Nothing but spider webs and dust."

We never went back to Dry Cave; the beauties of Endless Cave were so near and so entrancing that it just did not seem worthwhile to forego the delights we knew were there for the remote possibility of finding something interesting in this uninviting hole.

POSTSCRIPT

We did miss something at Dry Cave. Bill Burnet, Carlsbad paleontologist, made several trips into the cave when he heard about the bones, and identified a number of Pleistocene animals, including horse, bison, and camel. Our romance of the murder victim died with his reports, for he found no human bones.

In later years, particularly in the late 1960's and early 70's, interested cavers began trying for more cave, and at last report the cave is still going, with several thousand feet now recorded. The reports, however, indicate very little of actual beauty, with just dry crawlways and undecorated rooms predominating. Being so close to once beautiful McKittrick and Endless caves, the cave engenders hope that just through the next crawlway will be a Soda Straw Room or a Coral Room to beat anything ever seen in Endless.

Hamblen's Cave

"Hey, Sam," it was Dave's voice on the phone, "whatcha doing tomorrow?"

"Nothing special. Why?"

"Hamblen's got wind of a cave out in the mouth of Slaughter Canyon. He says it sounds like it may be a good one. You want to go look for it?"

"Foolish question. Tell him we'll pick him up at six o'clock."

So it was that by seven o'clock the next morning, April 29, 1934, the three of us were well on our way to Slaughter Canyon, having just passed the cluster of tourist cabins and the curio shop owned by Charley White at the entrance of Walnut Canyon and the road to Carlsbad Caverns.

"This is supposed to be the cave where old White gets the formations he sells to the tourists," Glenn was explaining.

"How come he told you where it is?" I asked. "I tried to get him to tell me, but he never would."

"Oh, Charley didn't tell me. I got to talking to a man that used to work for him, and he helped Charley haul some of the formations out of the cave once. Charley has since fired him, and he's mad about it, so he told me where the cave is."

"Do you think we can find it?" Dave asked. "Charley told me one time no one would ever find it, because he had the entrance covered up."

"I don't know, but this guy said it was quite a ways out from the entrance of the canyon, almost out in the flat country. He said to watch for a dim road that turns left off the canyon road about half a mile from the mouth of Slaughter. He said it would really be hard to see at first, as Charley tried to keep the first fifty or one hundred feet cleaned of tire tracks so no one could find it. If we find it, we are to follow it about a mile to within a hundred feet of a low hill with a bluff about twenty feet high. The cave is at the foot of this bluff, and Charley keeps a bunch of cactus stalks over the entrance."

"Sounds like a cinch," Dave commented.

But it did not prove to be. We could find no sign of the dim road turning off the main one. We left the car at about the right distance from Slaughter Canyon's entrance and fanned out on foot, sweeping the countryside to a line well south of the left slope, but we never saw a sign of a road. After a couple of hours of fruitless walking, we pulled up on the banks of a wash and sat down for a drink and to talk the situation over.

"Either that guy lied to you," Dave said, "or rains have washed out all signs of the road."

"So now what?" I asked.

"We might go up to Rainbow."

"Before we do that," Glenn put in, "let's follow the line of this wash past the low hills until we get clear into Slaughter. Maybe we're too far out."

The line of hills climbed gradually in gentle slopes to meet the first majestic upsurge of Slaughter Canyon's wall. We started walking toward the mountain. Dave edged out into the flatland clear of the hills, I moved in closer to the gravel wash that bordered them, and Glenn climbed up and followed their edge, to scan the ground below. We had not gone a hundred yards when Glenn shouted us to a stop.

"Here it is, right below me."

He had hardly skittered to the bottom of the slope before Dave and I joined him. There was a cave there, all right, almost hidden in a pile of boulders close up against the foot of the hill.

"I can't say that it looks like much," Dave complained.

"Neither does the entrance of Spider," I retorted. "This could be a dandy."

Glenn climbed down into the circular, cistern-like cavity that 89 dropped about four feet. A dark crawlway led off from the bottom of this, straight into the hillside. He lay down on his side and shone his light into the hole.

"Well . . . ?" Dave questioned.

"It goes on," Glenn said. "Looks pretty tight and bends out of sight in about ten feet."

"It's your cave, boy. Go see what you can find."

Glenn's feet had hardly disappeared into the tunnel before he came backing out, a lot faster than he had started in.

"Jeez but that was quick!" Dave exclaimed.

"There's a centipede in there as big as my arm."

"That would be one helluva centipede!"

"Well, anyway, pitch me a stick so I can defend myself."

We threw him a dead branch from a nearby shrub, and he started cautiously back into the hole. It seemed he would never crawl completely out of sight. Finally he called back, his voice muffled. "I don't see him. I guess he went into a crack. The tunnel gets larger just around the bend, so come on."

No sooner had I entered the passage than I noticed a not too ancient smell of skunk and the all too familiar acrid odor of bats.

"Dave," I called back over my shoulder, "if I meet a skunk, you'd better start making room, because I'm coming out."

Where the tunnel curved to the left it started sloping slightly downward and getting gradually larger. Finally we could proceed by stooping, and we then stepped into a small chamber about twelve feet square. Directly in the center of the ceiling hung one lonesome, brown stalactite about two feet long; a stumpy, gnome-shaped stalagmite stood on the floor beneath, seemingly with head cocked back looking up at its companion above, as though seeking the source of the dropping water that had formed it. That had been long ago, for the room was now as sere as Hell's kitchen, and the stalactite had started scaling off with dryness and age.

Chimney, Dry, and Hamblen's Caves

The tunnel continued out of the room, but it was a crawlway again. We inched along it on hands and knees, making a couple of right-angle turns as it changed course. In thirty feet it opened into another room nearly twice as large as the first one, even more barren, and with no central stalactite to break its monotony. A large pile of rubble had built up against the far wall, reaching almost to the ceiling. The hole we had crawled through was the only opening; there was no other way out.

"I'll bet the cave goes on behind that stack of rocks," Dave remarked, "but it sure would be a helluva job to move them."

His voice started a squeaking overhead and a whispered fluttering of wings. At the side of the room, where the ceiling joined the wall, a small colony of bats hung upside down, their tiny eyes sparkling in our lights. The whole mass swayed nervously as we played our beams over them, and one or two dropped loose, to wheel in erratic flight about the chamber. A pile of guano a couple of feet deep covered the floor beneath the roost; it was evidently a favorite haunt.

"Hey," Glenn called from the other side of the room. "This looks like writing scratched into the rock."

There *was* something there on the wall about shoulder high. About six lines of illegible figures stretched some three feet across the wall, but the characters were so old and indistinct that nothing could be made of them. Possibly they were merely natural fissures and cleavages in the rock, but the uniformity of the lines and the even height of the tracings gave them the appearance of having been drawn there.

"'Keeping time, time, time, in a sort of Runic rhyme,'" I quoted from a favorite poet.

"Glenn, let's get Sam outa here; he's going off his rocker."

After we scrambled back into daylight and sat down on the edge of the cistern for a drink of water, Dave remarked, "Well, this definitely wasn't Charlie White's cave. Let's call it Hamblen's Cave. After all, he discovered it, dared its dangers to explore it, and bravely led the way into it, in spite of a centipede as long as his arm."

"That's the story of my life," Glenn complained. "A chance to have a cave named after me, and what do I get? Sarcasm, a flock of bats, and an unknown message scratched on a wall."

If there was more cave behind the pile of tumbled rocks, and if that was a message carved on the chamber's wall, I'm afraid they are forever lost. In 1943 a sixteen-inch deluge pounded the mountains, and the flood waters poured from the mouth of every canyon in the Guadalupes. Slaughter Canyon washed from slope to slope in boiling torrents many feet deep. When the floods had run their course, a great plain of tumbled rocks, debris, and gravel covered the canyon floor, spreading out like a great white fan for several miles beyond its mouth. The spot where we found Hamblen's Cave was ground down and buried, and the course of the wash we had followed to find it changed completely.

I have no idea how to come even close to the spot where Hamblen's Cave lies buried.

11
Kindel Cave

"THERE," TED YELLED, EXCITEDLY, "I told you I could find it! That's the cave entrance right up there."

"You could lead us right to it," Dave jibed. "We've only been looking since nine o'clock and it's now almost three."

"Maybe so, but I got you here."

Five of us—Tommie Futch, Dave Wilson, Glenn Hamblen, Ted Fullerton, and I—had started out that cold morning of January 28, 1934, to find a cave that Ted had spotted several years before from the floor of Dark Canyon, deep in the Rattlesnake Bends area. He remembered only rather vaguely how to find the roads to get there, but did recall that the entrance was "high up on the steep slope, and big enough to fly an airplane into."

The entrance was big, all right, looming huge and black well up toward the canyon's rim. The steep climb up to it looked awesome. Tommie got a fire going and brewed up a pot of coffee, and we gulped down a sandwich apiece before starting the climb. I noticed Glenn ate only a bite or two; he looked pale around the gills.

"I don't know," he replied to my inquiry. "I think I got carsick bounding around on those damnable roads trying to get here. I'll be all right." But it turned out he was not; we had not climbed more than fifty feet before he gave up, "heaved his cookies," and headed back to the car. But he insisted that the rest of us go on to the cave and not worry about him.

The scramble up to the cave was a holy terror. The presence of long stretches of gravel slides did not improve the situation. Once Dave's feet shot out behind him, and he plummeted for thirty feet, belly down, hands grabbing for any support, cap and flashlight flying away, finally coming to a stop against a clump of prickly-pear cactus.

"Damn," he moaned, picking thorns from various portions of his anatomy. "It could have at least been a clump of shinnery."

"You'd better be thankful for the cactus," Tommie drawled. "That's a twenty-foot drop just below you."

Finally, we gained the entrance and clambered up to a flat, smooth shelf just under the overhead arch. At least 300 feet below us we could make out Glenn puttering around the fire, and we knew he was all right. Far out across the landscape one rolling range of hills drifted into another, and then another, until the far one fused softly with the blue haze of distance.

We noticed right away that the rocks of the entrance and sections of the canyon wall surrounding it were black with soot and that the passage behind extended into the darkness of the cave.

"I understand goats often use the cave for shelter during the winter," Ted explained. "Old Man Williams said he remembers when the goat manure caught fire and burned for days, blackening everything."

The next thing we noticed was three or four heavy anchor bolts firmly implanted in the arch of the entrance and a level, built-up roadbed leading back into the darkness. Several wooden rails, lying asunder, and cross-ties, warped and out of place, littered the roadbed. Ted filled us in on the history of that, as he had heard it. It seems that some time after the turn of the century, two men named Mudgett and Kindel had wandered into the cave and discovered a vast deposit of bat guano back about a hundred feet from the entrance. Realizing the value of this as a fertilizer, they set about developing mining operations for its removal. They made a half-hearted attempt to smooth the slope leading up to the cave's mouth, installed a windlass in the opening for letting the sacks of guano to the canyon floor, laid the roadbed and rails from the entrance back to the deposit, and somehow worried a heavy flatcar up and onto the rails.

Then, after all this work and expense, their plans for some reason fell through. Some say they discovered that the guano had become worthless through years of leaching and exposure to the dampness of the cave. Others say their finances ran out. Whatever the reason, they never realized much in return for their labors. But evidence of their efforts was still to be seen. We had noticed the windlass lying on the floor of the canyon, broken and bent. We were standing on the old rock roadbed. We found the old flatcar sixty feet back in the darkness, lying on its side and half buried in the guano. Fifty or more sacks of the stuff, once neatly stacked in piles, cluttered the floor; the fabric had rotted away, and the guano had poured out to join again the brown deposit from whence it had come.

"And that's the story as I heard it," Ted finished up.

"What about this Kindel fellow?" Tommie asked. "He must have been kin to Sonnie."

"I believe he was Sonnie's dad," Ted answered.

"Then let's give our old cave-exploring buddy due recognition by naming this cave after his father," Tommie suggested. "Besides, after all the work that went on here, I think the old man deserves it."

So we always called it Kindel Cave; in later years when someone talked about Mudgett Cave, we knew which hole they were referring to.

"Well," I said, "did we come here to explore a cave or talk about its history? Besides, this spot stinks to high heaven." It did not smell like a bed of roses, what with the odor of burned goat manure, the animal stink of the herds, and the acrid fumes from the guano.

The corridor stretched straight ahead into the mountain on an almost level course, but the going was by no means smooth. Large pits, not deep but treacherous, dropped off on each side of the corridor near the walls. The passage was about a dozen feet wide and the same high, but after about fifty feet it enlarged to almost double that size. There was a disappointing lack of cave decorations, with only a few straggly, two-inch stalactites, brown and unimposing, spotting the ceiling. We passed the overturned guano car and the sacked fertilizer, then made a sharp bend to the right into the room where most of the actual mining operation had taken place. Deep pits stared darkly from the deposit which covered the floor to an unknown depth. A rusting shovel, its handle splintered and broken, stuck upright in the center of the room. An old, rotten, blue denim jacket lay half-buried over near the wall, and the rusted remains of an ancient kerosene lantern still hung from a metal peg driven in the wall.

"You can almost sense the ghosts of things past in here," Tommie almost whispered, his flashlight probing the dark recesses.

We noticed two or three small passageways dropping off into darkness over against the left wall, but we continued down the main corridor, which swung to the left from the old mining room. We still encountered small patches of ancient guano, laid down by more venturesome bats, and commenced finding spots where water dripped from above. When this occurred above an old bat roost, one had to tread carefully, as Dave discovered to his discomfiture, because the ancient guano on the floor, once waterlogged, was treacherous as ice. And it was certainly a mess to struggle up from.

Occasional squat, knobby stalagmites began to appear here and there on the floor as the corridor narrowed and started sloping gradually upward. Then we stepped into a roughly circular room probably twenty feet across and ten feet high. In its center, completely dominating the scene, rose a mound of dark-brown formation six feet across, smooth and shining in some places, pleated and curtain-like in others. From its top rose two columns of chocolate brown, as thick through as a lad's body, fluted and terraced to the ceiling. The room was in deep contrast to what we had seen, and seemed beautiful in our lights; dripping water made the whole center group glisten.

91

"Now, this is more like a cave," I said. "I've got to have a picture here."

Dave suggested we call this the Room of the Black Temple, and no one offered anything better.

Just beyond the Black Temple the corridor sloped gradually upward, and the floor peaked at a huge column that seemed to bar further passage. It was twice as tall as a man and six feet through, and in appearance resembled a congealed flow of molten tan silica that had poured from a hole up where the wall joined the ceiling, to fall to the floor in a knobby columnar mass and spread out in a widening circle down the passage. Across from it a frozen cascade dropped in two lovely terraces down the wall and hung across an alcove like stone draperies.

92

We made our way around the Candy Waterfall, as Dave christened it, and found the corridor narrowing down and the floor leveling out. Over against the right wall hung a long row of marble curtains, a deep orange in color, dripping water and glistening in our lights. Some of the larger ones resembled elephant ears, and when we shone a light behind them, it was transmitted soft and honey-colored.

93

I found a small natural basin on a shelf about shoulder high. Dripping water from above kept it filled, and the overflow formed a paper-thin sheet of onyx that hung down from the edge like a gauze curtain. We drank from this tiny pool, and found the water pure and sweet.

"Boy," one of the fellows said, "I am glad you found that. I was getting thirstier'n hell, and that dung-filled stuff we've been finding just didn't appeal."

We found several tiny tunnels leading off at floor level. We had to lie prone to shine our lights into them, but our efforts paid off because we discovered a veritable forest of mini-stalactites and helictites, all glistening with water, ranging from a pure white to a deep chocolate brown. Small pools of water reflected the beauty of these fairy passages, a beauty that could be viewed only from without, for only a fairy could walk them.

The corridor narrowed, the formations became fewer and disappeared altogether, and in a short distance we were crawling down a tunnel that emerged in a few yards into a small, circular room from which no passage led.

We had reached the cave's end.

"And I guess it's about time," Ted mused. "It's getting late outside, the road's long and rough to town, and Glenn's probably getting lonesome."

We stepped out on the ledge at the cave's mouth to a twilight just deepening to dusk. The far hills were purple, the nearer ones each giving way to a lighter-tinged one behind. A pale pink suffused the sky at the horizon, drifting softly upward to fade away into the deep blue bowl of the heavens. We could see a fire blazing down by the car, and knew Glenn still held forth in his solitude.

"Sure hope he's got the coffee hot," Dave shivered in the winter cold creeping down from the crags. It was faster, but not easier, slipping and sliding down the mountain slope. The darkness did not help, but we made it without more serious injury than a banged-up knee or a scratched hand. Glenn was well again.

Over three and a half years were to pass before I made another trip to Kindel Cave. We were well into August of 1937, and my last year at college was just around the corner. Tommie and I had never been certain we had seen all of the cave on our first trip, so we decided

94

that before I left for school we would get in one more underground jaunt to see if more could be found in Kindel.

Nothing much had changed, except we found the cave a lot drier than before. Just out of reach of daylight we were examining a small passage that promised to lead somewhere.

"Listen," Tommie said. "I heard something over there behind us."

At first I heard nothing; then a distinct scratching sound came from the vicinity of the far wall, and a soft, muted whisper, like heavy breathing.

"There's something over there, all right," I replied.

"Maybe one of those ghosts we were talking about last time."

We eased over to the opposite wall, and found a five-foot depression beneath an overhang. A large billy goat lay there in the low cavity; as our lights picked up his image, he tossed his head, and his horns scraped the ceiling. His eyes glowed yellow in the darkness, the black oblong pupils in sharp contrast, a certain frantic wildness in their expression.

"Well, I'll be damned," Tommie said. "He must be sick, or crippled."

The words were hardly uttered before Billy came unwound. With a snort and one terrific bound he cleared the alcove and almost knocked us sprawling as he tore between us and broke for daylight, bat guano and gravel flying.

"I guess there wasn't anything wrong with him," Tommie remarked dryly.

We carefully worked the cave to the crawlway and the small circular room at the end. Several branching tunnels invited exploration, but none of them amounted to anything. Only one of them gave promise of leading anywhere, and we were just getting excited about it when it emerged back into the main corridor after a distance of thirty feet.

Our lost basin up on the ledge, from which we had drunk our fill the last trip, was dry, and we were getting thirsty. Tommie glimpsed a suggestion of moisture high on the wall near the ceiling, and managed to clamber up a fissure to investigate. A small ledge was there, at the back of which dropped a tiny, deep crevice. Tommie stretched an arm down into the hole.

"I can reach water," he grunted. "Hand me a dipper."

"How about an empty film carton?" I asked. It was the only thing I had that might hold water. We each managed to get several good swigs before the glue dissolved and the cardboard carton disintegrated.

Although we found no new cave to explore, re-visiting the half-remembered old scenes was rewarding. As we stepped out into the entrance, the hot breath of August swirled about us. The distant hills rolled away one after another, shimmering in ethereal dance as the heat waves boiled about them. We left the cave behind us, silent and somber.

12
Lake Cave

WHILE HOME FOR THE summer vacation from college in 1937, I heard of a recently discovered cave high up on the north slope of the right fork of Slaughter Canyon. It reportedly was away up near the rim, and held a great lake in its depths. When I told Tommie Futch about this feature, he was as positive as I that we had to see it. I finally got more precise directions for finding the cave, so on the evening of June 23, after we both got off work, we set out to find it. The rocky terror that posed as a road up the canyon had not improved since I had last traveled it, and we soon left the car and headed up the north fork slope.

Almost miraculously we found the cave, just as the dying sun edged the tips of the far cliffs with gold. Our discovery was even more remarkable because the small opening, only about ten feet wide by half as high, nestled at the base of a bluff at the back of a narrow ledge. From ten feet away the hole was completely hidden.

"I guess St. Christopher was with us tonight," Tommie said. "I still don't know how we found it."

"But find her we did. So let's see what she's got."

Stooping to get under the low arch of the entrance, we found ourselves in a room about twenty-five feet wide and twice as long, with the ceiling a dozen feet overhead. We slipped and slid downward for the first dozen feet; then the floor leveled off, progressed smoothly through several large, irregularly shaped stalagmites, and pitched off almost vertically for another fifteen feet. The drop here was not difficult, as it had plenty of handholds, and we found ourselves in a great chamber that rapidly expanded to a width of over a hundred feet. The ceiling seemed fifty feet above us, and the 250-foot beam of our lights disappeared into the blackness ahead.

"Quite a cave we've got here . . . no?" Tommie asked.

Many white and light-tan stalagmites clustered about, and they took on weird, nightmarish shapes in the dim light. Over in the shadows near the wall a prehistoric dinosaur sat on his haunches and glared at us. A giant rhinoceros poked his horned muzzle from behind a white hillock. Out in the center of the room a Pope's mitre soared to a height of eighteen feet. We walked behind the rhino and looked back, and it became a great dragon, wings spread, rearing thirty feet into the air.

"This is a menagerie of monsters," Tommie commented, as we wandered along. "Look at that giant toad over there against the wall."

Standing about among the monsters were occasional graceful, tapering spires like Siamese temples. A continuous line of stalagmites, with turreted walls, some even having windows, stretched at one point for nearly fifty feet, splitting the great room. Over against the wall behind the Pope's Mitre a massive flowstone cascaded for twenty feet, merging on the right side with a row of terraced spires rising fifteen feet into the air; from one angle the entire mass resembled a castle, and we called it "Castle Rock."

"There's the old crone I used to have for a neighbor," Tommie shouted, indicating a stalagmite out in the center of the room. "She could put out more gossip over the back fence in an hour than the scandal sheet could print in a week." And from the side, this stalagmite, standing twice as high as Tommie, was almost a perfect replica of the head an old woman, hair in a tight knot on top, nose tilted in the air, mouth open and loudly expounding on the carryings-on of those young folks down the street.

Just beyond the Gossiper a ragged stalagmite tapered thirty feet toward the ceiling, like a shell-torn steeple in a war zone. At one side of the room a great, gaping hole fifty feet across plunged for fifteen feet. Its sides were covered with beautiful terraces of circular travertine, resembling lily pads turned to stone. This had definitely been a great pool of water some time in the past, but was now dry.

After wandering enthralled for probably 300 feet, we came to a jumping-off place, for this great room dropped off into another whose floor continued thirty feet below us. The rim of the drop was edged with a fluted dam standing about six inches high and never more than 102 an inch thick, which wound about in graceful contours for many feet. We could see that at one time it had contained the waters of a pool that had stretched back a dozen feet from the edge of the drop. At one point a beautifully tapered stalagmite sprang up for ten feet, like a huge Chinese temple set in a Lilliputian wall of China.

"This cave has the damndest things," Tommie muttered, as he searched for an easy way into the room below.

We found a fissure sloping down, and, half clambering, half sliding, we skittered to the lower level. On the way we saw the beams from our lights glisten on water ahead, bouncing reflections up among white stalactites hanging from the ceiling. Soon we stood on the brink of the largest pool we had ever found in a cave. Squatty brown stalagmites resembling blanket-clad Indians gathered about in the darkness.

"It's a lake!" I exclaimed. "A real underground lake."

We estimated it to be at least thirty feet wide, as it almost completely spanned this new room, and the beams of our lights were dim at its far end, at least a hundred feet ahead. We could see many spots that were definitely ten feet deep. The water was crystal clear, and the bottom and sides of the pool were covered with a deep-brown, spongy-looking growth of calcite. Scattered about among the sponge-heads lay a large number of white stalactites that had fallen from above; they gleamed white in our lights in the clear water.

"This is really something," I said. "No name but Lake Cave will do for this one."

Three stalagmites grew up from the lake floor in different places. One of these, over six feet tall, beautifully terraced and tapered, we called the Lighthouse, for there was nothing it resembled more, towering there above the lake. Another one we called the Pointing Hand, 103 for it looked like an outstretched hand, fingers together, standing white and pure on a bulging pedestal of fluted brown. An old waterline showed clearly around the lake, fully two feet above the present water level. Everything above the old waterline gleamed pure white; everything below glowed a soft rich brown.

"Tommie!" I exclaimed. "Let's take a swim. I've just got to say I've been swimming in an underground lake."

"Not me," he retorted. "That water's probably cold as ice. A guy'd be a damned fool to jump into that."

"Then I'm going to be a damned fool. And just to prove I did it, I'm going to set up for a picture and you open the shutter and fire the powder after I get in."

"I'll do 'er, Sam."

I stripped off my clothes, felt the chill air of the cavern on my bare skin, took a look at the clear water, and wondered if I *were* a damned fool. But not giving myself a chance to lose my nerve, I plunged into one of the deep holes. The water was so cold I think I almost bounced right out of it in shock. I know I heard my yell echoing in the great room.

"What's the matter, Sam? Is she a little cool?"

"Cold! Gawdalmighty, this is the coldest stuff I ever got into."

"You wanta wait for the picture?"

"I'll wait," I replied, teeth chattering, "but if you see me turning blue, throw me the rope."

By the time he got the picture taken, I am certain I was turning blue. But when I tried to scramble up the rough, spongy, hide-torturing shoreline, I was thankful for a little numbness of feeling; that stuff would have been murder on a fully sensitive skin.

The Room of the Lake was only about half the size of the large one with the fantastic stalagmites that we had just visited, but it was beautiful beyond compare. Great masses of honey-colored flowstone rippled down the walls, and white stalactites covered the ceiling. At the right side of the lake a dark-brown stone waterfall cascaded from a crack at the ceiling line, covered the wall in a mass twenty feet wide, and plunged into the water. At one point just above the shoreline a slender white stalagmite no bigger around than my arm connected floor and ceiling, forming a fragile, fluted column thirty feet high.

At one side of the room a narrow hole led down for a dozen feet into a small room whose ceiling was a mass of brown and white stalactites. As I stooped to enter the chamber, my head struck and broke a stalactite about the size of an old-fashioned candle, and a stream of water poured for many seconds from the break.

"I never saw that happen before," Tommie marveled, watching the water stream out, as though from an open faucet. "I'm going to try that again."

He reached up and snapped another stalactite off, and got another stream of water, but it did not pour as much or last as long as the first one.

We returned to the lake room, and up over a ledge near the back we found another long, low chamber, almost like a balcony. Its ceiling was only about four feet high, and the little room was only about six feet wide and twenty feet long, but it was a charming miniature forest of slender columns of white. It seemed that almost every stalactite had joined the floor, and there were thousands of them, so closely spaced a mouse would have had difficulty winding his way among them.

At another place back in the Room of the Lake a great brown mass of flowstone, shimmering in delicate draperies, covered the wall for a space of a dozen feet. In a corner stood a huge stalagmitic group that resembled a feudal castle, with turrets, towers, great gates, and even windows. Nearby, a stalagmite reared fifteen feet into the air, then spread out at the top in a great circular lip, its underside deeply fluted, like a giant toadstool. A deep purplish brown coloring the gills of the underside added to the illusion.

105

We found no more rooms or corridors, but we really did not need to, for there was already so much to see, so many creatures pictured in the grotesque stalagmites of the first room, so many strange, unearthly objects, so much sheer beauty, that we lingered until the last batteries for our lights grew weak and we were forced to leave for fear of being caught in the darkness.

The pre-dawn air was dry, pungent, and heady. The spicy odors of the desert mountain flora welled up from everywhere. As we stumbled down the last few feet of steep canyon slope to the floor of the canyon, we heard, off to our left only a short distance away, the soft, plaintive "Poor Will, Poor Will," of a nighthawk. It came again, but now it seemed to our right, down in the brush at the bottom. "Poor Will, Poor Will." This time it certainly came from behind and above us.

"As intangible as a ghost, that feller," Tommie muttered softly. "And just as lonely."

13

Lodgepole Cave, McKittrick Cave, and Rainbow Cave

Lodgepole Cave

IN THE SPRING OF 1939, when Bill Burnet, Seth McCollom and I visited Gunsight Cave high on the north slope of the canyon of that name, we spotted the entrance of a cave that looked promising. As we walked the high ridge about a mile west of Gunsight Cave, we saw the black entrance like an ink spot down on a projecting shoulder of the canyon. The climb down to it threatened to be a rough one, but I determined that some day I would go to it.

It was almost a year later, February 18, 1940, that Tommie Futch, J. D. "Static" Burke, and I found ourselves on the high ridge, looking across at the mysterious black hole and trying to decide the best way down to it.

"I think we ought to back up," Static suggested, "to where the shoulder starts out, then go down it to the cave."

"I don't know," Tommie considered. "That's a pretty steep drop. And some of those ledges won't be easy; they're a lot higher than they look."

"It seems to me," I put in, "that the best way would be to drop straight off here to about the cave's level, then circle around to the right, following that ledge that goes just above it."

"You might be right," Static agreed. "I don't believe it's as steep."

"I'm with Sam," Tommie said. "Let's give it a whirl."

It turned out there was no best way down; it was a rough go from the moment we left the ridge crest. Ledges that had seemed to be only steps from the top developed into muscle-grinding drops of ten or more feet, and the slopes between were so steep that we negotiated most of them by the slip and slide process. We finally gained what we thought was the crucial ledge that led to the cave; this was pure guesswork, because we had lost sight of the opening shortly after dropping off the ridge.

The ledge turned out to be anything but a paved highway, but at least it offered horizontal scrambling. After what seemed a mile of this, we stopped for a breather.

"Fellows, we've missed the damned thing," Tommie said, examining our back track. "We're well around the shoulder, and the cave was on the near side."

"I think we are still too high up," Static disagreed. "I believe it's just below that next ledge down."

We dropped down to the shelf below and started back in the direction from which we had come. Within a hundred yards just a corner of the opening appeared at the base of the ledge we were following, about thirty feet below us. The small portion visible from our position was black and promised great things to come.

We found a cleft in the bluff and scrambled down to the entrance level. The hole seemed disappointingly small and was almost hidden behind brush and cactus.

"Guys," Static said, "we've found the wrong cave. This isn't the hole we saw from away up there."

"No, but this one is," Tommie called out. He had prospected around a bulge in the cliff and was out of sight to the left. We joined him, and there was the hole we had seen, large and black and brooding.

"You could see this one from up there, all right," I remarked.

"I'm glad I brought this," Static said, starting to uncoil the rope from his shoulders. Tommie slipped down to the lip of the entrance and peered in.

107

"I don't think you'll need it," he called out. "Someone before us has already taken care of that."

We joined him to see what he meant. We saw a crude, makeshift ladder, a twenty-foot length of pine tree with the branches lopped off, leaving six-inch stubs. The top rested against the upper edge of the vertical drop into the cave, while the bottom stood firmly on the floor beneath.

108

"Reckon she's sound?" Tommie wondered, giving it a testing shake or two. "Seems solid enough."

"All the same, I'm going to string the rope down beside it, just in case," Static put in. He tied one end of the rope around an oak shrub just outside the entrance and tossed the coils in.

"There you are, Tommie," he said. "You found the main entrance first, so you get the honors."

"Sam's the one who showed it to us. I think he ought to lead the way."

"I gladly relinquish that right to you, Tom," I bowed, with exaggerated courtesy.

"Pikers will be pikers," he came back. "Just give me room."

He wrapped the safety rope around one arm and stepped off onto the log ladder. About halfway down he paused and bounced up and down a couple of times.

"Sound as a dollar," he commented, clambering on down to the cave's floor. "There's quite a hole here; come on down."

We joined him, finding the ladder steady as a rock and surprisingly easy to negotiate. While we were getting down the log, Tommie had been exploring about.

"Fellows," he said, "I think we've got one of those things."

"A big one?"

"A big nothing. All I've found is this one big room with the small entrance up around to the right. If she goes anywhere from here, it's through a small hole I've overlooked."

He had overlooked nothing. We were in a big U-shaped chamber, some fifty feet in its longest dimension, with a ceiling fifteen feet above. The room was dominated by a great fluted column almost in the center. It was so huge that the back side had grown solidly to the wall behind, forming the U. At the left leg of the U was the entrance with the log ladder; at the opposite end was the smaller entrance we had first seen. Most of the walls were plated with flowstone, smooth and dry; to the right of the column the flowstone projected in a serrated knob, like an erosion remnant, that resembled the skull of a giant cat. The floor was covered with breakdown, and only a small, lonely stalactite or two adorned the ceiling. Not a drop of water appeared anywhere. Not a single tunnel branched off inviting exploration.

"Well, I think we've had it," Tommie commented, flashing his light about.

109

"Hmmph," Static grunted. "It's more like we'd been had."

I set up a flash picture to show the giant column, the only truly photogenic thing in the cave, and the smoke from the powder followed us up the log ladder.

"Well, what shall we name her?" Static asked, as he coiled his rope.

"How about Almost Cave?" Tommie suggested. "She was almost a cave."

"I think she deserves better than that," I put in, "if for no other reason than the stiff climb she put us through to reach her. How about Lodgepole Cave for the pine tree ladder?"

"I don't rightly think that's a lodgepole pine," Tommie drawled. "More like a ponderosa."

"I like Lodgepole," Static said. "Sounds kinda western and Indiany. Who knows, maybe the Apaches put that log in there many years ago."

So Lodgepole Cave it became. We took a long swig of water from our canteens and faced the slope before us. We knew that what appeared to be the ridgetop far above us was just an illusion created by the perspective of the downward slope, and that many false horizons would appear above us before we reached the true one. Many were the curses we heaped upon the innocent cave that had tempted us down that tortuous slope.

"She just wasn't worth it," Static gasped as we finally staggered onto the crest of the ridge and faced the two-mile walk back to the car.

At the moment I felt inclined to agree, but now in retrospect, with the toil softened by the haze of years, I know that the muscle-racking climb down to and up from the cave serves as one of the many memories needed to fill out my treasury of delightful caving experiences.

McKittrick Cave

From the standpoint of human history, McKittrick Cave ranks as one of the oldest caves in the Guadalupes. A date as early as 1894 was once seen on its walls, and by the 1900's it was a favorite spot for young couples on a day's outing from Carlsbad, then the small hamlet of Eddy. The cave lies only about twenty air miles west of Carlsbad, but by horse and buggy it took an entire day to reach the cave, have a

110

picnic lunch at the entrance, put in a few hours exploring—and, regrettably, removing formations as souvenirs—and return to town. The ride home probably was often made by moonlight, a circumstance intentionally arranged by the romantically inclined young swains in the groups.

111

In one of my mother's old albums I found a large number of photographs made on picnic excursions to McKittrick Cave. They would have to date at about the turn of the century. Many were taken by flashpowder inside the cave, and are remarkably clear. Considering the freedom and sensibleness with which today's cavers dress, one has to be amused at the attire of the cave explorers in these old photographs—the men in white shirts, stiff collars, unfurled cravats, and frequently sporting vests and straw hats; the women in flowing skirts that swept the ground, high-collared blouses with puffed sleeves, now and then even a perky chapeau balancing atop a bun of piled-up hair.

112, 113
114, 115
116, 117

My first trip into McKittrick Cave was on March 29, 1926. I was still working for Ray V. Davis, and he had often mentioned a desire to visit the cave, having heard so much about it as a former popular picnic area. He had hopes of taking a few photographs as a historical record.

"Why don't we take off tomorrow and go see it?" he asked.

"Suits me fine," I replied. My desire to see the cave had been whetted by my mother's early photographs and the tales of the good times she had there.

"I'll arrange for a lunch. I suppose we should take only the 5 × 7; if I understand correctly, the cave is probably too small to drag the 8 × 10 around in."

Neither of us knew how to get to the cave, having only a general idea of its location. But we struck out, and after a couple of wrong roads, some advice from a friendly rancher riding by on his horse, a flat tire, and getting stuck on a high middle—a common hazard on the rutted roads of the backcountry—we found the cave.

"Not too impressive an entrance," Ray remarked as we walked up to the opening. It was about the size of two washtubs and dropped down for about eight feet. A rickety old wooden ladder was lying off to one side. We tested it and decided it would probably suffice for the short drop into the cave.

"Even if it breaks, the fall won't kill us," Ray concluded, as we

Lodgepole, McKittrick, and Rainbow Caves   137

placed the ancient structure in the hole. Several fragments of shattered stalactites lay about in the entrance, certain evidence of the vandalism we had heard about.

We had hardly passed the extreme edge of daylight before we realized that it would have been a mistake to bring the bulky 8 × 10 camera with us. By the time we reached the first chamber large enough to be called a room, we decided that even the 5 × 7 was unnecessary. Evidences of the great beauty that had once reigned here were everywhere, in the form of fractured stubs of stalagmites on the floor and broken nubs along the ceiling where delicate stalactites had once hung by the hundreds. Fragments of stalactites littered the floor, as though a warped mind with a giant hand had swept them down to hear them shatter on the rocks.

"Nothing to photograph here," Ray said, a tinge of sadness in his voice.

We plodded on, from corridor to room, and room to corridor, and the scourge of vandalism was everywhere. True, we did find occasional spots where some formations had been left intact; in a cave apparently once so beautiful, it would have been almost impossible for a few not to have survived. At one point we found a really beautiful stalagmite standing practically unscathed amid the wreckage. It stood about ten feet tall, white and graceful, its sides decorated in fluted draperies between smooth terraces, the whole tapering to a graceful spire. Amost two feet thick at the base, it had probably proven to be just too weighty for anyone to attempt removing. One of the flat terraces, four inches broad, near the base had been sanded smooth by human hands, then polished to a mirror finish.

"Must be the work of the old hermit I heard lives somewhere around here," Ray remarked. "I understand he used to make tombstones out of the bigger formations."

As we left the room, I glanced back. The white stalagmite stood stark and alone. I could not help thinking of a painting back in a barber shop in Carlsbad. It pictured General Custer standing alone and defiant, his soldiers dead at his feet, the instruments of his inevitable destruction closing in from all sides. We left the cave without taking a single photograph.

A year later, March 3, 1929, a group of twelve of us, looking for a

place to picnic, decided to go to the cave. The party consisted of Jay
and Bertha Leck, Ruby Lee Jones, Lawrence and Forestina Matthews   118
and their two small children Larry and Latina, Grandma and Grandpa
Matthews, E. F. Brunneman, Harold Renfroe and myself. I had warned
them of the destruction they would find, so they were not too disap-
pointed; those who had never gone cave exploring enjoyed themselves.
The chamber where General Custer made his last stand was barren
and deserted. The General was gone, and only broken remnants of his
army lay about. Someone had devised a way to remove him from the
cave.

That was the last trip I made to McKittrick Cave. We were in the   119, 120
immediate area the day we discovered beautiful Endless Cave in the   121, 122
spring of 1934. I had seen the awful vandalism of the old cave, and
here was a virgin one of endless beauty; nothing called me back to
McKittrick.

Rainbow Cave

"It's one of the most beautiful caves I ever saw," Carl Livingston told
me. "You won't believe the colors you'll find there. Pinks, and reds,
and yellows, and greens, even delicate shades of blue."

"How could all those colors get in a cave?" I asked.

"They're caused by algae, which grow only in lots of bright
daylight. The huge entrance of this cave provides plenty of that."

We were exploring the high country to the south of Slaughter   123
Canyon, searching for Indian petroglyphs, or virgin caves, or anything else that wild, precipitous land might offer. We had surprised a
herd of blacktail deer, and a short time later Carl pointed out the
tracks of a mountain lion in a sandy wash beneath a steep bluff.

"When we reach that high point up yonder," he said, pointing to
his right, "I think I can show you about where the entrance of the cave
is. You might want to go there some day."

When we reached the point, the vast awesome gash of Slaughter
Canyon dropped off beneath us. Far to the west the blue heights of the
Guadalupes faded softly into a late afternoon haze. Straight ahead,
and far across, the north slope of the canyon reared up in pitching
slopes and sheer reddish walls hundreds of feet high.

"Right over there is the Kneeling Elephant," said Carl, pointing to a narrow, vertical knife-like upthrust that swept up toward the canyon's rim. "From the road leading to the canyon's mouth it looks just like the outline of an elephant."

He pointed out the spot to the right of the Kneeling Elephant where we would find Rainbow Cave. "If you're careful, you can't miss it, it's so large. But you have to walk right up on it before you can see it, because it drops off right at the base of a high cliff."

On a cold, drizzly day, March 15, 1931, my brother B. A. (he was *never* called by anything but his initials!), Harold Renfroe, and I set off to find Rainbow Cave. The rain was still falling, almost a fine mist, when we stopped the car just beyond the entrance of Slaughter Canyon. There had been no mistaking the Kneeling Elephant as we approached the canyon; now it loomed overhead.

"You mean we have to climb clear up to there?" Harold asked, shielding his eyes from the rain. He was only in his middle teens, and I am certain his first view of this wild country had left him in awe.

"Clear up to there," I answered.

"But before we do," B. A. put in, "we're going to have to have something to eat." We found a sheltering overhang which protected us from the rain, and B. A. got a fire going. Soon he had bacon sizzling and was stirring up a pan of scrambled eggs.

By the time we finished eating, the rain had quit falling, and we started the arduous climb. The rocks were wet and slick, and Harold spent almost as much time down as he did up. Even B. A. and I hit the ground a couple of times. Harold may have been a poor mountaineer, but he was the first to discover the cave.

"Here it is!" he shouted from a short distance over to our left. "Here it is. My gosh, come look at this hole!"

It *was* quite a hole. The opening was like a right-angled triangle lying on its side, the apex to the right, the base line running for about thirty feet to the left, with the high point of the triangle some twenty feet above. The hole was smack up against the base of a cliff; if we had been fifty feet to our right we probably would not have seen it.

As we stood in the entrance and let our eyes become accustomed to the dim light in the cave, we slowly began to realize what a really tremendous hole opened up before us. The ceiling of the great chamber

124

extended straight ahead, level and smooth, until, far back just at the edge of darkness, it began a gentle slope downward. The floor started dropping away right at the entrance and continued sloping at an acute angle until it disappeared into the shadows at the far end. The room widened at the right just within the entrance, and seemed to continue at a uniform width of about eighty feet.

"What a hole," Harold repeated, as though he could not believe what he saw.

As we made our cautious way down the slope, we became more and more aware of the display of color. Overhead the ceiling was a strong pink that became more and more delicate, to shade gradually into a pastel blue at the back. The right wall was buried beneath a cascading flowstone waterfall colored a deep velvet green. Almost directly across on the opposite wall a similar cascade spanned floor to ceiling and extended for over fifty feet, and it was tinted a vivid orange. A great column bordered this on the left, stretching in tan and orange splendor entirely to the ceiling, which seemed at least a hundred feet above. The column had grown so close to the wall that for more than half its entire length they were firmly joined.

"No wonder Carl Livingston called it Rainbow Cave," I mused. We were about halfway down the slope, when Harold shouted out a new surprise.

"Hey, there's a skull here, stuck in the top of this stalagmite."

We hurried over to join him, and were struck by the similarity between the odd configuration of the top of the formation and a flattened human skull. The stalagmite stood about four feet high, knobby and irregular, with rough popcorn outgrowths thick along the back side. The death's head faced the cave entrance. The eye holes were narrow, evil slots instead of round cavities, and the grinning teeth were well outlined above the jaw line. The skull sutures were sharply defined, and the flaring nostril holes gaped above the protruding teeth. Algae had colored it a deep moss green, giving it an ancient, opened-grave appearance.

"Sure looks like something out of a nightmare," B. A. remarked.

We passed a great mounded stalagmite connected to the right wall that stretched upward for eighty feet. About two-thirds of the way up it was bisected by a notch eight feet high, creating a sort of canopied throne; above the throne the formation rose in three rounded sections, ending in a circular dome still several feet beneath the ceiling. The left side of this formation toward the cave's back and the entire front were colored a deep green, while the side facing the entrance was splashed with an overcoating of white and orange, as though someone had sloshed several cans of paint down its sides.

"I remember reading a fairy tale," Harold said, "in which the princess sat on a canopied throne and waved her magic wand. We ought to call that the Canopied Green Throne."

About two-thirds of the way back into the big auditorium, the ceiling started sloping downward at a fairly steep angle to meet the floor at the back. Here stood the cave's most massive stalagmite, as though barring the way. It was an irregularly shaped thing protruding here in rough, knobby bulges, sinking there in cavities and crevices. It lacked any symmetry whatsoever; in several places it seemed to have changed its mind as to whether to grow out or up. About twenty feet through at the base, it towered over halfway to the ceiling, which we estimated to be well over a hundred feet above. A connecting wall of flowstone, twenty feet high, joined it to the left side of the cave, forming a huge giant's saddle. The stalagmite was a deep orange-brown in color, but it was so splotched over with green algae that the original color was more than half buried.

We made our way around and behind the giant, and as we turned to look back at it, we were struck by the fantastic silhouette it made against the strong light from the entrance and the smooth ceiling far overhead.

"Looks just like a giant with upraised club standing guard over his underground world," B. A. said.

We reached the back of the cave about 150 feet behind the Guardian, where the ceiling sloped down to meet the floor. Everything was twilight here, and the vivid colors had faded to faint pastels or an overall bluish grey. Only a faint blush of pink remained on the ceiling.

"Gee," Harold lamented, "do you think this is all of the cave?"

"There's a hole up there above that ledge," B. A. pointed out. "I'll crawl up there and see if it goes anywhere." He worked his way up to the ledge and disappeared. In a few minutes his head came into view, and he called down, "It goes into a room that's not very big, but there may be a passage leading off."

"Wait for us," I called.

We joined him and entered a small room, about twelve feet square, colored all over a soft delicate blue. A few short stalactites decorated the ceiling, and ribbons of flowstone, like tiny rivulets of frozen ice, streamed down the walls. One miniature pool of water caught a drip from a stalactite above. It was the only water we had found in the cave, and there was barely enough to quench our thirsts. If the tunnel B. A. had seen led anywhere, it pinched down too tightly for us to have any desire to explore it.

"I suppose we should call this the Blue Room," I said as we stepped out on the ledge and the great auditorium stretched out before us. As we scrambled down to the main floor of the cave, I noticed a narrow crevice off to one side and at the very back of the room.

"That might go somewhere," I suggested. We squirmed into it for about twenty feet. Once it widened out and promised to lead somewhere; then we faced a breakdown of scrambled rocks.

"And that's all there is," Harold said.

As we made our way up the steep floor toward the glare of light from the entrance, B. A. stopped and looked about. "How come I didn't notice these as we came in?" he asked, pointing to numerous squat, round-domed stalagmites that stood about like gnomes. They were scattered all about, ranging up to about four feet high. The sides facing the entrance were fairly smooth and colored a deep green. The back sides were coated with white popcorn-like protuberances resembling underwater coral. From the side, the white in contrast with the dark green made us think of white-haired dwarfs, or, as B. A. commented, "Like Uncle Toms in a church choir."

As we stepped out of the hole into the outside world, we felt the exhilaration of the rain-washed air, heavy with the smells of wet cactus and mountain shrubs. The mist was gone, and it seemed a giant hand had torn great rents in the grey canopy overhead. A shaft of pure golden sunlight streamed through one of these, far up the canyon where the air was still heavy with moisture. It burned like a celestial searchlight, its rays falling squarely upon a high-flung cliff, turning it to burnished bronze against the soft purple ridges behind.

"Well," B. A. asked Harold, "what did you think of your first shot at cave exploring?"

"I'm ready to go again!" he exclaimed.

It was to be almost three years before I again visited the great cave at the foot of the Kneeling Elephant. I had been haunted with the thought that such a big cave must have more than that one great chamber, and had never lost the desire to go back and look for more cave. So, on January 14, 1934, Tommie Futch, Dave Wilson, Glenn Hamblen, and I decided to go find the rest of the cave, if there was any.

We carefully inspected the side walls as we went down. Dave even crawled up to look behind the Canopied Green Throne, and paused for a moment while I took his picture there. At the far back of the cave, where there was quite a bit of breakdown and cracks split the walls, we searched diligently, but found only a few fissures that led nowhere. Glenn crawled into the small tunnel leading off from the Blue Room, but soon backed out.

"It might go somewhere," he told us, "but it sure would be a tight squeeze. Sam, I don't think you could even get in it."

We scrambled back down to the cave's floor, with the black outline of the Guardian looming ahead.

"Who needs more cave?" Dave asked, hands on hips, gazing about the huge auditorium. "We've got enough right here."

14
Painted Grotto, Crystal Cave, and Sitting Bull Falls Cave

The Painted Grotto

THE PAINTED GROTTO REALLY CANNOT be classified as a cave, but its distinctive character and unique decorations place it securely in a special niche in any discussion of the caves of the Guadalupes. It actually is only a great overhang, or shelter. The only darkness any part of it experiences is the darkness of night, for it extends back into the mountain to a depth not greater than twenty-five feet. Overhead, the graceful bend of the arch peaks at about twenty-five feet and spans some fifty feet in its curving swoop to the floor, which is level and benchlike.

On this day, February 18, 1934, Tommie Futch, Dave Wilson, and I were taking a cousin of Tommie's on a cave-hunting excursion. This cousin, nicknamed "Dough" Spell, who was visiting from Mississippi, had become intrigued by our stories of cave exploring and insisted on being taken into a wild cave.

"How about us going up Slaughter Canyon and having a look at that big hole we spotted away up on the south slope, the day we went to Goat Cave?" I suggested.

"I don't know how much of a wild cave it will be," Dave put in, "but it'll be a wild enough climb up to it."

The road up the canyon was not too bad, considering the terrain it crossed; the rancher who ran stock in the area had watering places scattered about and maintained the road to a certain degree in order to service them. When we stopped the car, however, the ruts were dim and tortuous and the rocks had taken over. We pointed out to Dough the cave we proposed taking him to; even at its far height up near the rim, the opening made a deep, dark gap in the side of the mountain.

"'Pears like a mighty fur piece," he joked. "I hope these Mississippi legs hold up."

"Say," Tommie interjected, "if I'm not mistaken, that cave with the Indian paintings is just a little ways on up the canyon. Why don't we have a look for it first?"

"The Painted Grotto!" Dave exclaimed. "I've heard of it and sure would like to see it."

"Let's see if we can find it," Dough agreed.

We rounded a bend only a short distance beyond, and the big arch of the shelter loomed right ahead of us on the left, not fifty feet above the canyon's floor.

"I wish they were all as easy as this," said Tommie, laughing, as we mounted the easy slope to the opening. When we gained the ledge-like floor beneath the arch, we paused for a moment to let our eyes adjust from the bright sunlight to the deep shadow of the overhang. At first we saw nothing; then strangely fantastic figures and designs began taking shape all along the back wall.

"Just look at them!" Tommie exclaimed. "They cover the wall."

He spoke without exaggeration, for they did exactly that. In fact, in many instances designs lapped over onto other figures or completely covered older symbols beneath. Almost every color had been used, but a dark, rust-colored red predominated. We saw a large number of greens and yellows. The wall was a maze of color as far up as the Indian artists could reach. In fact, many designs were painted eight to ten feet above the floor.

"Must have been a damned tall Indian that painted that deer up there," Tommie quipped, pointing to the stylized outline well above his head.

"They probably piled up rocks and stood on them," I said. There were plenty of them scattered about on the floor and tumbled down the slope beyond.

We began trying to see how many figures and symbols we could identify. Some of them were easy—no mistaking the saucy roadrunner flirting his long tail to the skies, the jagged streaks of lightning flicking down from above, the long-tailed lizard almost obliterated by a running deer added by a different artist at a later date.

"Look here," Dough called. "This has got to be a man raising a heavy war club. I don't know if that's a deer or a dog he's bashing."

"Boy, some turkey left a lot of tracks here," Dave commented, pointing out a line of them weaving their way across some older symbols.

133

"Hey, this looks like a saddle."

"This must be a cloud with the rain falling out."

"Look at this shield in two colors, outlined in white."

"If St. Peter were here, he could really be playing this harp."

"That's gotta be the longest-legged bird I ever saw."

"I wonder if this could be the product of a dirty mind."

Many of the designs were purely symbolic. Long rows of hash marks in deep red marched across the wall, amid daisy-chains of white circles and crazy zig-zagging lines of orange and yellow. Dominating the entire show was a horizontal, dark green serpent fully twenty feet long, painted with careless abandon atop any design or figure unfortunate enough to have been drawn previously. The entire body of the snake was outlined in white triangles about two inches high.

"Must have been a rattler," Tommie observed, indicating the constricted left end of the figure, with the white triangles much smaller in size and in two parallel lines close together.

"Wouldn't it be something to know what every one of these things meant to the old Indians?" Dough pondered, down on his knees examining a kite-like emblem near the floor.

"I wouldn't be surprised if a whole lot of it wasn't just plain doodling," Dave interjected.

We could have spent hours amid these fascinating denizens of a distant past and of unfathomed minds, but our true goal was an unknown cave high on the jagged canyon wall above us, and we would need those hours to achieve it.

As Tommie put it, "Let's leave these ghosts to their trials and pleasures, and get on with the job at hand."

Crystal Cave

It was a good thing we had the hours, because that climb from the Painted Grotto up to the big, black hole not far beneath the canyon's rim was a time-consuming, lung-stretching, muscle-straining pull. Loose rocks and gravel slides and precipitous drops threatened at every turn. Cacti of all shapes and descriptions, thorny and merciless, lurked by every rock and bush. In most places the ascent was so steep and rugged that fifty feet of progress without resting was quite an accomplishment.

"I gotta stop and get my breath," Dave gasped at one point. We were by then well up on the mountain, and the opposite rim now approached eye level. Dave was carrying a sotol stalk as a staff to aid in climbing, and he paused on a rocky ledge that overhung a gravel slide which began about eight feet below; it plunged precipitously down a narrow ravine. I had stopped about fifty feet away from Dave, beside a large boulder; sharply etched on my mind is the picture of his figure, in silhouette against the sunlit canyon wall beyond, leaning on his sotol staff and looking up at the steep climb remaining. I did not like the looks of the ledge he was on; from my angle it seemed crumbly and insecure.

"Dave," I called out, "I think you'd better get off . . ." Before I could finish my warning the rocks crumbled beneath him and he spun off, arms and legs flailing, amid the tumbling rocks, onto the gravel slide below. I remember how high his staff went into the air. Somehow he landed on his feet, and he and the flying rocks from the disintegrated ledge and the uneasy gravel underfoot started a headlong plunge down the funneling ravine. I realized with a start that the whole thing ended a hundred feet below, where the ravine pitched off a precipice of undetermined height.

"Dave!" I yelled, powerless to assist. Twice he went to his knees in the landslide, but managed each time to bounce erect. Then, somehow, he sensed the precipice ahead and realized his danger. How he managed it I'll never know, but in two great bounds he cleared the sliding mass, falling flat on his belly, to clutch the branches of a shrub growing on the edge of the ravine.

Another picture is sharply etched on my memory: Dave's white face turned up at me, his expression so woebegone it would have been laughable under other circumstances, his glasses dangling by one earpiece, his prematurely bald pate shining in the sun, and the cloud of dust swirling up over the lip of the fall where the gravel poured over.

"Well, David . . ." Somehow, it was all I could say.

Amazingly he came through that experience with nothing more serious than deep scratches and abrasions. But I think if we had not been just below the cave we sought, he would have given up for the day and started back down. We knew we were near the cave, because Tommie and Dough had climbed a shoulder to our left and called out that they could see the entrance just above them.

I do not believe any cave anywhere has a more beautifully symmetrical entrance than this one. It forms a wide, low oblong, with smoothly rounded corners at the top, as evenly proportioned as though designed by a draftsman. And its proportions dwarfed us into insignificance. We estimated its height at twenty-six feet, and stepped off its width at ninety-seven feet. The inside of the arch was stained a beautiful, delicate pink.

"Reckon our Indian friends painted that?" Dough joked, looking up at the roof.

"Carl Livingston told me the beautiful colors in Rainbow Cave were caused by algae," I replied, "so I guess the same holds true here."

We discovered a series of confusing corridors branching off from the entrance chamber, but a brief exploration revealed that they all ran together again a short distance within the cave. From that point the corridor, about twelve feet square, extended straight into the mountain. A few white formations decorated the passage. Within thirty feet we emerged into a roughly circular room some eighteen feet in diameter. The beauty of this room made us catch our breaths, for it seemed we had entered a snow palace. The walls were white, covered with frosty flowstone like frozen waterfalls. Snow-white stalactites hung down in masses from the ceiling, and stalagmites just as white reached up toward them from the floor. The absence of color in this gleaming white room set it apart from all we had ever seen.

Two corridors branched off from the Snow Palace. Feeling they probably came together as had those at the entrance, Tommie and Dough started down the right one, while Dave and I turned toward the left. We had taken only a few steps when we heard Tommie call.

"Sam, you and Dave come here. We've found the Torture Chamber."

That certainly had intriguing possibilities, so Dave and I hurried to join the other two. The passage they had entered emerged within a few feet into a small room, probably ten by twenty feet, with the ceiling only about seven feet above at its highest point. Tommie was right, for this room certainly resembled a spike-filled chamber of horrors. Walls, ceiling, even the floor, were covered with great crystals of the calcite known as dog-tooth spar. They protruded in every conceivable direction, ranging from small ones only an inch or two long to huge ones up to ten inches in length and six inches through at the base. The big ones predominated, and they presented an almost fearsome appearance, bristling like hundreds of spikes in a chamber of the Inquisition. It was almost as though we were inside a huge, crystal-lined geode.

Unfortunately, the crystals were all encrusted with a coating, a fraction of an inch thick, of lusterless calcite. When we broke one off, however, the interior gleamed like a great transparent jewel.

"What a sight this would be if only the crystals weren't coated," Dave lamented.

The corridor Dave and I had started down did not amount to anything, and we realized we had seen the cave.

"Can't think of a better name than Crystal Cave for this one," Dave said as we emerged into the great entrance chamber.

Dough was enthralled by the panoramic view through the wide window of the entrance, with a precipitous bluff intruding directly to the right and the vast sweep of the canyon dropping straight down. Far below gleamed the white ribbon that marked the boulder-strewn course of the flood waters that occasionally surged through when cloudbursts drenched the slopes. Far across, almost at our level, the opposite wall of the canyon shimmered softly in an afternoon haze that had drifted in.

"Nothing like this in Mississippi," Dough murmured. "Nothing at all."

Sitting Bull Falls Cave

About thirty miles slightly southwest of Carlsbad, as the buzzard flies, Mother Nature has staged a scene so completely incongruous to the surrounding terrain that the beholder can hardly believe that what he sees is really there. The road winds for about five miles through a desert canyon in the first reaches of the Guadalupes. As one penetrates the canyon, the barren walls soar higher and higher, grim and lowering, with only the unfriendly cacti breaking the monotony of the sere crags. It is called Last Chance Canyon.

Then the road turns to the left up a small side canyon a few hundred yards to a parking area. A short walk leads around a projecting shoulder of the canyon, and there it is, gentle and delicate, beautiful and unbelievable. Far up at the right rim a hundred feet or more, a wisp of water plunges over a moss-covered cliff in a vertical drop of twenty feet, cascades down a pitching slope to a bulging overhang another twenty feet below, then drops sheer to the rocks and a waiting pool some sixty feet beneath. If a wind blows in the canyon, as it often does, as much of the falling water whips away in spray as falls to the catch basin.

A little stream drains from the brim of this pool and winds beneath a green tree and through reed-shadowed ponds, to cascade down a gentle slope to the canyon floor. Minnows dart in the pools, butterflies and dragonflies flit about, and frogs can be rousted from the reeds and grass to plunk wetly into the ponds. Lemon-yellow columbines nod in Narcissan fantasy along the water's edge or play hide-and-seek among the rocks and shade in the falling mist. Such a lush scene and tumbling waters just do not belong at the end of a canyon so arid and bare.

Just as strange, behind the cascade of falling water hides a beautiful little cave, dripping, lake-filled, and delightful. It has derived its name from the one given the falls, Sitting Bull Falls. How it acquired such a name does not belong in a story of caves, but suffice it to say that the great Hunkpapa Sioux medicine man was never in the canyon, never saw the falls, and bears no responsibility for their name.

My first visit to the scene was in the mid 1920's, over a road so rough and terrifying I still recall it with dismay and remember the three flat tires we fixed on Davis's old Chalmers touring car. At that time a series of footholds had been cut into the solid rock leading up behind the falling water. Some believe these had been carved by Indians using the hidden alcove as a religious place, but early ranchers denied this and claimed white men cut them there. Anyway, Kenny

Davis and I climbed up the steps, took a thorough drenching in the falling water, and stopped just on the edge of darkness, for we carried no lights. The cave loomed ahead, black and mysterious. A lake of water covered the floor, and we tossed stones out into it and as far back as we could, trying to determine its extent. A swim in the darkling waters seemed the only chance for exploration, and we had no inclination for that.

It was not until April 19, 1931, that I returned to the beautiful little canyon and its waterfall. This time I was in the company of Carl Livingston, member of a pioneer ranching family, whose interest in the history of the region had led him to write of its past and of the beauties to be found in its interior. Many of his articles appeared in eastern periodicals and in the old *Wide World* magazine, published in England. His goal on this trip was to make some flash powder photographs in the cave behind the falls. A drenching shower was inevitable for anyone visiting the cave, as the way up by the carved footholds led directly under the falling water. Carl protected his camera and powder in an old yellow cowboy raincoat.

"Brrrr," he shivered, when we gained the alcove. "Should have worn the slicker."

His lantern was stubborn, but he finally got it lighted, and in its glare we made out the details of the rather large room we were in. Water dripped everywhere, and beautiful tan flowstone covered the walls. A maze of stalactites hung from above, and several massive columns, brown and glistening, soared to the ceiling fifteen feet overhead. All but a small section of the floor was covered with a lake, but we could skirt it by following the wall.

"The first time I was in here, many years ago," Carl related, "the water level was much higher—you can see the old watermark up here on the wall. The other fellow and I—I believe it was one of the Ward boys—carried sotol torches, and we waded into the lake to try and see how far back the cave went. We tossed stones ahead of us to determine how deep the water was; we could tell by the plink or the plunk if it was shallow enough to wade, or might require swimming. It was quite an experience."

We made our way to the far extreme of the lake and paused on a small incline that served to impound the waters. I thought nothing could be prettier for a picture, and suggested it to Carl.

"This is great, all right," he answered, "but I remember a small room on a lower level I'm very anxious to get a picture of. It was once a lake, also, but most of the water was gone, and the walls were just like a big dry sea bed of massive coral. I want to take the first picture there, because I've a feeling the smoke from the flash powder won't allow another picture, and it may run us out of the cave altogether."

We found the room, just beyond and to one side. The passage sloped sharply down into it, and everything was encrusted with a sponge-like growth which did resemble coral. The chamber was rectangular, with a beautifully decorated ceiling a dozen feet overhead. At the far end a sizable remnant of the old lake still lingered, but most of the room was dry. All around the wall, on a line well above our heads, the old water level was plainly evident. Everything above the water line gleamed white and soft tan, while, in startling contrast, everything below was a dark reddish brown. The coral heads, some twice as large as a man's body, were in the dark zone, protruding from the walls in irregular masses extending to the floor. Where stalactites had reached the old water line, they ended in bulging blobs of the coral-like growth.

"This is it!" Carl exclaimed excitedly. "This is where I want my picture!"

He poured out a pile of powder up near the entrance to the sunken room, placed about two feet of dynamite fuse in it, checked to be certain I was properly placed in the scene and that his camera shutter was open, and then put a match to the fuse. He scrambled down to join me, keeping his flashlight shaded and turned away from the camera. We waited and waited. Nothing happened. He had cautioned that we had to remain perfectly still so there would be no blurred movement when the powder ignited; we sat like frozen statues.

139

"Well, by George," Carl muttered, squirming about to look toward the flash, "I don't believe the powder is . . ." Before he could finish, a flare of light burst into the room with a blinding boom. In the small confines of the chamber, it seemed to me the stalactites rattled.

"Got it," Carl exulted, scrambling up to snap the camera closed.

Smoke swirled about us, and we hurried to the upper level to escape its choking fumes. But it followed us into the main chamber, and we were forced to leave the cave.

"Gee," Carl worried, "I wonder if I used too much powder and burned out the picture."

"You certainly made enough smoke," I retorted.

When we developed the film the next day, we found he had not used too much powder. The negative was fine.

In the winter of 1931-1932, a prolonged period of extremely cold weather settled in on the mountains, and ice started forming on the falls. It built up and built up into great masses that bulged off the face of the cliff where water poured, and huge icicles weighing hundreds of pounds reached for the canyon floor. Eventually the weight became too much, and the entire overhanging face of the wall, where the main waterfall began, tore loose and crashed to the ground below, smashing into hundreds of boulders, some weighing tons. The cliff face here was not too stable, anyway, since it was a travertine deposit many feet in thickness which had been built up by centuries of falling water and dashing spray.

The shearing away of the cliff face behind the falls left the first room of the cave almost half exposed, and obliterated the old hand-hewn footholds leading up behind the falls. But time, the great healer, with the aid of perpetually falling water, has softened the scars of the catastrophe, and visitors today to the scene are not even aware that the great mass of fallen boulders was once a part of the cliff above.

POSTSCRIPT

I made another trip into the cave behind the falls on August 25, 1946. The area then, as now, was supervised by the Forest Service. A paved road led to within a hundred yards of the falls. A fine parking area had been set aside, and numerous picnic tables were scattered about, with cooking hearths and protecting roofs.

My wife, Elizabeth, was with me, along with her brother, Bob Fleming, and we were taking our three-year-old son, Aaron, on his first caving trip. He did not think much of it, and spent most of his

138

140

time underground wanting out. I found the cave much drier, with most of the entrance lake gone. The inevitable signs of vandalism darkened my memories of past beauty. Hardly a stalactite that could be broken off remained above, and many stubs along the floor marked the former sites of gleaming brown stalagmites. Even the coral heads in the sunken chamber bore evidence of hammering, but most were too massive to budge.

In spite of this destruction, the cave still maintained a sort of ethereal, enchanted beauty all its own. Perhaps it was the scramble up the boulders and along a trail cut by the Forest Service beneath the falling waters, which still drenched us with an unwanted shower, that set the stage for the unexpected charm waiting just beyond in the darkness.

I would not be surprised if this place really was sacred to the Indians of long ago. Maybe they *did* cut those footholds in the living rock to reach that charming and mysterious sanctuary.

15
Whistle Cave, Tunnel Cave, and Wind Cave

Whistle Cave

DURING THE SUMMER OF 1936, I spent the vacation period between college semesters working for the Cavern Supply Company at Carlsbad Caverns. One evening Tommie Futch came by our home.

"Sam," he asked, "how'd you like to go on a treasure hunt?"

"I've gone on several of those," I answered, "and I'm still poor as Job's turkey. I'd rather go cave exploring."

"Well, this will include some of that."

"O.K. I know you've got something up your sleeve. Spill it."

"Well, I've got this map here," he drawled, pulling a piece of folded paper from his pocket and spreading it out. "Don't ask me where I got it, and I don't know how good it is. But you notice it pinpoints a cave up the south fork of Slaughter Canyon. I'm sure it's the one we spotted up beyond the Gash the day we went to Crystal Cave. If not, it's close."

"So . . .?"

"The guy that drew me this map is positive there's a bandit's treasure buried in this cave. Some distant cousin, or somebody, told him about it and sent the original of this map. He's too old and crippled to try to get to the cave, so he's given me the map for a fifty-fifty split of whatever we find."

"I don't have much faith in lost-treasure stories," I replied, "but if there's a cave involved, let's go."

So it was that my next day off found us bumping up the would-be road into the south fork of Slaughter Canyon. The "thoroughfare" up the canyon badly needed maintenance; it may not be the world's worst road, but it certainly comes close. We were forced to a halt a short distance from Painted Grotto, and headed on up the canyon on foot, carrying a pick and a shovel, and various lanterns, flashlights, camera, and lunch bags.

About a mile up the canyon, we rounded a bend and came in plain view of the cave Tommie had in mind. It was rather high up on a gentle-seeming slope, and made a brooding, black arch in the face of the mountain. Tommie checked his map.

"That's gotta be it," he exclaimed. "There's the Gash over there, and here it is on the map. The canyon works out just the same, and that's the slope the map shows leading up to the treasure cave."

I admit to a little excitement as we re-shouldered our tools and began the long trudge up to the cave. It really was not a bad climb compared to the one up to Rainbow Cave, or Crystal, and we soon stood in the entrance, peering down into darkness. Brush grew thick and heavy around the opening; we had to force our way through. The entrance arch was some twenty feet overhead, spanning a breadth of perhaps thirty feet.

"Let's go get our gold!" Tommie shouted.

The cave's mouth opened into a rather large auditorium, the floor of which sloped sharply down. It was covered with broken rocks and gravel slides, and we made our way down with difficulty. Once my feet went out from under me, and I covered several yards in a horizontal position. The shovel I carried went flying, and clattered almost to the floor of the chamber.

As I got to my feet, Tommie started to make some remark about my clumsiness, then stopped, held up his hand, and said, "Listen . . ."

Up from the depths of the cave, out of the darkness and the unknown, came an eerie, pulsating whistle that fluctuated, loud, then soft, then died away, only to come on strong again, wavering and shrill. I felt the hairs stand up on the back of my neck, and a rippling chill went down my spine.

"What in the hell . . .?" I started to explode, but Tommie stopped me with raised hand. It came again, tremulous and haunting.

"Whatever it is, it isn't alive," Tommie muttered. "Let's go see what gives."

We scrambled to the floor of the cave, which still sloped gradually down into the darkness, and made our way back into the mountain. The strange whistling filled the air about us, then died almost to a whisper, only to swell out again. It grew stronger as we penetrated farther into the blackness.

Then we were at the back of the great room. A blank wall faced us, curving up to the ceiling a dozen feet overhead. The whistling was all about us. It seemed to come from the walls, some thirty feet away on either side, then from a point up near the ceiling; once I was certain it came from down near the floor. While I was investigating this possibility, and making up my mind to quit this haunted place if we didn't find an answer soon, Tommie moved up against the back wall, cocked an ear up along a shallow fissure, and then scrambled up a few feet to an eye-level position before a narrow ledge.

"Well, I'll be damned!" he exclaimed. "Come take a look at this."

I worked my way up beside him, and could not believe what I saw. Someone had taken an ordinary double-disc metal whistle, commonly found in dime stores or Cracker Jacks, and firmly cemented it into a cavity which originally must have been about six inches across. A strong draft of air pulsated in and out, formerly through the cavity and now through the whistle, creating the haunting whistling sound that drifted eerily about in the great room as the air oscillated back and forth through the tiny metal toy.

"Someone with a peculiar sense of humor has been here before us," I remarked.

"Hey, maybe that's the key to our treasure," Tommie said.

We examined the wall carefully all about the whistle for some sign, but there was nothing else. But it seemed like as good a site as any, so we set to with spade and pick, digging quite a cavity at the base of the wall beneath the whistle. It would be great to report on the treasure chest we uncovered, filled with gold coins and antique jewelry; but, like most seekers of golden troves, we found only sore backs, blistered hands, and sweaty brows. If a treasure is buried there, it moulders elsewhere than beneath the whistle.

There really was not much to see in the cave besides the great chamber sloping down from the entrance, which bore very few decorations. A few scraggly stalactites an inch or so long clustered here and there on the ceiling, and a few small folds of draperies made a

141

half-hearted attempt to decorate the walls, then gave up. One stalagmite, rather graceful and encrusted with a coral-like growth, lorded it over all at the back of the room, near the whistle. It was all of eight feet tall, and was the only true stalagmite in the entire cave.

"Well," Tommie mocked, as we started the weary trudge up to the entrance, "we ain't got no treasure, and we ain't got much cave."

"But we've got one hell of a whistle."

As we stopped for breath beneath the entrance arch, I noticed some splashes of red over on the west wall. We investigated and found a dozen or so old Indian pictographs painted on the rough rock with red ochre. There were turkey tracks and parallel stripes and vertical bars enclosing dots, a triangle, and a running deer. One design seemed to be a human figure, completely stylized, bowing at the waist, arms dangling.

The whistle welled up from the depths, haunting and mournful.

"If I didn't know better," I muttered, "I'd say that was the ghost of the old Indian who painted these here, warning us away from his favorite haunt."

Small wonder that we named it Whistle Cave.

Tunnel Cave

I RAN INTO TOMMIE FUTCH and Sonnie Kindel drinking coffee at one of the drug stores, and joined them for a cup of brew. I had come home for the Christmas holidays, and had spent most of the vacation working on a term paper that would be due almost as soon as I returned to my studies at Missouri University.

"Do you guys realize I have been here ten days, and we haven't gone on a cave trip yet?" I asked.

"The last time I asked you, you were all involved in that thesis, or whatever it is you're writing." Tommie retorted. "I figured you'd holler when you were ready."

"Well, I'm ready, and if we don't make it this weekend there just won't be one. I've got to start back to Missouri about Tuesday."

"Well, I've got the cave," Sonnie interrupted. "A guy at the mine told me about it; he ran into it while deer hunting. It's up in Slaughter Canyon, and he claims it runs clear through the mountain."

"Now, that would be some cave," Tommie snorted.

"He probably meant through some ridge. It's up the left fork of Slaughter, pretty high up. As near as I can gather it's somewhere east of Crystal Cave, because he remembered that big hole."

"Well, then," Tommie said, "let's go find it. I just gotta see a hole through a mountain."

So sunrise that Sunday—December 30, 1936—found us bumping along the rocky ruts that led into Slaughter Canyon. The road got worse after we took the left turn into South Fork, and I was glad when Sonnie finally decided we had gone far enough.

"I think we ought to head right up that ridge there that parallels the one going up to Crystal," he said. "The guy described it as a big entrance, so we ought to spot it from somewhere up there."

"Sounds like as good an idea as any," Tommie commented. "Let's have at it."

A climb anywhere up the pitching slopes of Slaughter Canyon is a breath-grabbing, muscle-straining struggle. We were soon panting and puffing, with stops every fifty feet while our lungs gulped for air. Over across the canyon we could see the great vertical cave opening we called the Gash—not much of a cave, but a spectacular entrance. Far ahead, into the depths of the canyon but high on a shoulder, sat the graceful arch of Whistle Cave. We knew the Painted Grotto was below us and only a short distance up the canyon, and as we climbed we caught occasional glimpses of the perfect flat arch leading into Crystal Cave. The winter air was crisp, but the sun, burning just above the eastern rim, promised warmth, and the stiff climb gave it. We were glad we had left our heavy coats in the car.

After an hour's battle with the tortuous slope, sliding gravel, loose rocks, and ubiquitous cactus, we pulled up on a high spur, like a giant causeway protruding from the mountain.

"We damned sure ought to be high enough," Tommie panted. "I think Crystal is just around that shoulder."

I scrambled up a small incline onto a flat ledge and turned to face the rugged and majestic canyon. It tumbled off into the depths, the far reaches of which still lay in the early morning shadows beyond the reach of the mounting sun. Then I looked to my right.

"We're high enough, all right," I called out to the others. "Here's the cave right over here."

The near side of the next ridge over still lay in shadow, but the great entrance of the cave was so large it stood out, as Tommie said,

145 "like a sore thumb." It looked teasingly near, but a careful survey revealed a tough trip still ahead before we could reach it, with one deep, rough canyon to be crossed.

"By golly," Tommie said, "that looks like a cliff dwelling back in that entrance."

It really did, but when we resorted to binoculars for a better check, we found it was only a turreted wall of rock back in the opening. It took us another hour to gain the entrance of the cave. We found ourselves under quite an arch, for it swept across fully thirty feet above us and extended on each side for a span of more than sixty feet.

The cave did go through the mountain; far ahead a circular patch of daylight marked a back door out of the darkness. Actually, we discovered later that the ridge here was just a high, jutting shoulder sticking out from the mountain proper, and that the shoulder was not very thick. The cave penetrated it entirely in only about 300 feet.

Beyond the great entrance arch the cave consisted of just one big auditorium, its greatest dimensions being somewhat larger than the opening itself. This huge chamber dwindled gradually as it approached the small back door, becoming a corridor for the last fifty feet, then shrinking to the five-foot passage that exited on the other side of the ridge.

Standing just inside the main entrance was a pair of stalagmites
146 we named Mutt and Jeff. The taller one, about eighteen feet high, resembled a huge war club, while its companion, strikingly similar, was only about half as tall. Behind them, beyond a mound of flowstone, stood another about six feet tall that resembled a closed-hood toadstool.

"That's a beauty over there," Sonnie said, pointing over to the right where a massive column seemed to support the ceiling. Over twenty feet thick at the base, it was still an impressive twelve feet
147 through where it joined the ceiling. Terrace upon terrace it mounted up, topped for the last five feet with a graceful cascade of frozen limestone. It was a light brown in color, interspersed with sections of deep rust, with here and there a topping of pure white.

Tommie discovered a small tunnel pitching down at a sharp angle near the bottom of the great column. While I set up for a picture of Sonnie here, Tommie squirmed through the narrow opening, and we could hear him sliding off down a steep, gravel-strewn passage, finally to pass beyond the range of our hearing.

I had finished taking pictures, and we were beginning to wonder about Tommie, when we heard him coming back, again to the accompaniment of the crashing of sliding gravel; I believe we heard a curse or two.

"Well, what did you find?" I asked as he finally poked his head up through the opening.

"Not much that I could get to," he replied. "But there sure could be one hell of a cave down there. The tunnel got pretty large as I went down. Then, in about seventy-five feet it choked up with rocks and gravel. But I could feel a draft of air coming up through the rocks."

"Maybe we'll dig her out some time," I suggested.

"That would be one hell of a dig," he sputtered. "I don't think I want any part of it."

As we headed for the back door, we heard the squeaking of bats overhead, and our lights revealed a long narrow crack high up in the ceiling, almost spanning the auditorium. We could make out the figures of hundreds of the creatures hanging in the aperture, closely packed together. We tossed some rocks up toward them, causing a few to dislodge and fly about, but our actions resulted chiefly in an increase of the chattering and squeaking. It seemed strange that they had chosen to hibernate here in this great hall, with daylight filtering about them, and the cold winds of winter whipping beneath them between the two entrances.

We went out the back door, pausing to look back and admire the black silhouette of the great flat-topped arch of the front door, the rugged beauty of the canyon in the sunlight beyond.

"We might as well call it Tunnel Cave," Sonnie suggested. "Unless you guys can come up with something more appropriate."

Although we picked a new way down, it was not any easier than our original route up. With weariness nagging at our bones, we came to believe it was even harder.

We finally made it down to the car. A chill, biting wind ripped down the canyon, and we felt thankful that our heavy coats were there. Old Man Winter was making threats. Tommie stirred up a cheerful fire in the protection of a small bluff, and brewed up a pot of his southern hospitality coffee. That, with a few sandwiches left in our lunch bags, bolstered our spirits for the rough, bone-wrenching drive out of the canyon.

Wind Cave

We actually heard Wind Cave before we saw it. The date was May 6, 1934. Tommie Futch, Dave Wilson, Glenn Hamblen, and I were prowling the foothills a dozen or so miles southwest of Carlsbad looking for a cave on the old Judkins ranch. Old man Judkins had told us, rather vaguely, that there were several holes in the ridges on his place; all of them were pretty small, but he thought some might be good caves. We were looking for just any one of these.

We had spread out down the slope of a ridge, cactus-covered and rocky. A bare mile or so ahead of us the foothills merged with the plains that swept to the Pecos River, flared out to the east, climbed the caprock, and merged with the Llano Estacado, or Staked Plains, named by the Spanish Conquistadors before the Pilgrims landed at Plymouth Rock. Now, the flatlands danced in heat waves, and we could hear the shrill whirring of cicadas in the catclaw. The ocatillos were in bloom, punctuating the unfriendly slopes with exclamation marks of crimson.

I was following the crest of the ridge, while the others walked below me at intervals of fifty feet. I heard a sound that for a second I mistook for the air rushing through the wings of a dive-bombing nighthawk. As I signaled the others to stop, I realized this was a blast of air rushing through a narrow aperture; it actually seemed to hiss, like a steam vent. The others heard it, too.

"What do you think it is?" Glenn called out from down in the shallow canyon.

"It's definitely wind blowing from somewhere," Tommie answered from just below. "It's right down here ahead of me."

We followed the noise, and within fifty feet came to a large, flat rock lying on a level, open space along the sloping crest of the ridge. The air was hissing out from beneath this rock.

"What do you know about that?" Dave exclaimed. "Someone's hidden a cave."

148

As we leaned over to shove the rock aside, we could feel the draft, cool in the hot sun. It was no job for the four of us to tip the flat slab over. A small, round hole, about half the size of a sewer manhole, dropped straight down through the rock.

"Sam," Dave jibed at me, "if there's a cave down there, it's one you aren't going to see, 'cause you're just not going to fit in that hole."

The removal of the rock had stopped the hissing sound of the draft pouring out of the hole, but we could feel its force. Glenn picked up a twig and attempted to pitch it into the void, but it soared up about four feet into the air and dropped to one side. I was mopping the sweat from my face, and tossed my handkerchief at the cavity; it performed a parachute maneuver in reverse and I grabbed it as it passed over my head.

"Must be some hole down there for all that air to come from," Tommie remarked. "Reckon we can go see where she blows?"

"All but Sam," Dave mocked. "He'll never make 'er."

"I'm not that much bigger than you, you Louisiana swamp hound," I retorted. "If you make it, I'll be right behind."

I came to regret my boast. We could see that the shaft bottomed out about twelve feet down, seeming to open into a very small chamber. Glenn, being the smallest, was nominated to make the drop first. He could have seemed happier about the choice, but we assured him that if he found cave below he would soon have company. He eased himself into the hole, and the escaping air hissed around him. He dropped easily, checking himself against the walls, and we heard him thud as he reached bottom. He hunkered down and shone his flashlight about.

"There's some cave here, all right," he called up. "Not very big, but enough for all of us. Seems to be a larger tunnel through here to the right."

The others eased into the opening and slid down to join Glenn; Dave had to grunt and squirm a couple of times, but finally dropped clear.

"A dollar says Sam'll never make it," I heard him say to the others.

I started into the hole with my arms at my sides, and realized at once that just was not the way to do it, so I kicked myself out to free my hands, extended my arms over my head, and dropped into the hole. Only I did not drop; just as my eyes passed the entrance edge I squished to a stop. I squirmed a bit, but nothing gave.

"I told you," I heard Dave hoot below me. "Stuck tighter'n a cork in a bung hole!"

I could feel the contours of the hole against me, and somehow knew if I could just turn a little, I might get loose. I did a strangled hulahula, kicked a foot, and felt something rip, but I was free, and slipped a few feet before sticking again. But the hole had enlarged slightly, and I knew I had it licked; a few squirms and a reverse wriggle, and I fell to the floor among them.

"I want my dollar," I panted at Dave.

"You didn't call the bet," he answered, welching.

We duck-walked through a small passage to our right and emerged into a corridor that offered head-room and little else. It played out a short distance to our right, so we turned down it in the other direction. Soon we were stooping through a narrow, undecorated corridor, with the draft of air strong in our faces.

This was no prize cave.

Occasionally we could proceed upright for a step or two, but were soon forced to crouch or follow a tortuous passage on hands and knees. A few discouraged, stubby stalactites a couple of inches long hung here and there, and in a spot or two a thin ribbon of flowstone, like bacon, followed a crack in the ceiling. There just was nothing pretty anywhere, and I felt glad when the corridor finally pinched down, and the only thing that could get through was the rushing air.

I recall two or three small crawlways branching off as we made our way back to the room beneath the entrance, but no one seemed inclined to do that type of exploring. I know I had a vague worry about getting up the vertical squeeze to outside.

"You know," Tommie remarked, as we gathered for the climb out, "I don't believe the draft is as strong as it was when we came in."

"I thought down there a ways it had almost died out," Glenn replied.

"It does seem to vary," I put in. I pulled my handkerchief out and tossed it up; it settled slowly, fluttering slightly, to the floor.

"And I was depending on that wind to help blow Sam out," Dave remarked.

"You're going to be that wind," I retorted, "because I'm going out ahead of you. If I get stuck you're going to give me a push from below. And if I don't get out, you don't get out."

Glenn scrambled up first, followed by Tommie. "Keep your arms up," he called down to me, "so I can grab you and give a heave as she narrows down."

I wriggled my way up into the hole. Everything went smoothly for a few feet, then my belt buckle caught on something, and I had to squirm back a few inches to release it. Dave gave a push, and I moved on up, hearing my shirt rip at one tight spot. Finally I reached the constriction where I had hung up before, and everything stopped. I just couldn't move on. But Dave was puffing just below me, and Tommie reached my hands. With a push and a yank they popped me out like a stopper from a bottle, leaving most of my shirt and several patches of hide behind.

We called the hole Wind Cave. Dave wanted to call it Samstopper, but I just could not go for that.

In recent years smaller and more agile cavers have found a lot more to Wind Cave than we discovered. Stories have come out of delicate rooms of great beauty and variety, small but well decorated, of aragonite crystals and rimstone pools and of a River of Blood formed of deep red flowstone. No large chambers have been found, and much of the way is by stooping or crawling, but the cave goes on.

I have seen none of this. If I had difficulty making the drop into the cave when I weighed a mere 195, I would have to be a dreamer to try squeezing my present 225 pounds through. Wind Cave is just no longer for me.

16
A "New" Cave

"WELL, I DON'T KNOW," Charlie Stevens muttered, stirring his coffee. "I talked to one of the guys on that newspaper trip, and he said it was nothing but a dirty little bat hole."

"Yeah, I've heard some conflicting stories about that so-called exploration party," Tommie Futch interjected. "I really don't think Bill Burnet would have stated it equaled Carlsbad Cavern in beauty if there wasn't something pretty good there."

"Well," I said, "there sure is one way to find out. We haven't been in a cave in too long; let's go see for ourselves."

About a week later—September 18,1938 to be exact—dawn found us bumping along the ranch road that led to the mouth of Slaughter Canyon. After we passed the Colwell ranch in the entrance of the canyon, the road became only a reasonable facsimile, but fortunately the cave was only about a mile up the canyon, "high up on the south side" according to Bill Burnet. We stopped the car where some empty cans and an old campfire indicated others had halted before us.

"There's the big square rock Bill said marked the start of the trail up the mountain," Tommie indicated. "A right nice slope," he muttered, cocking his eye almost straight up the steep face of the canyon wall. "Bill said we couldn't miss signs of the trail going to the cave."

"A right nice slope" certainly was a gross understatement; we found it to be a breath-grabbing haul up, with loose stones slipping underfoot, often forcing us to hands and knees. Bringing up the rear, I finally staggered around a pile of rocks against the base of a cliff and found the other two sitting in a dark hole leading down into the mountain. We had made it.

"No wonder no one ever found it until Tom Tucker's lost goats led him to it," Charlie said. "With that pile of rocks in front, I don't believe the hole can be seen even from across the canyon."

"When I went to Rainbow Cave over there across the canyon we examined this side with field glasses, and I know it can't be seen from there," I replied.

A cool, faintly musty air blew up from the dark hole, which was roughly circular, about eight feet in diameter, in the base of a shallow alcove hollowed out of the cliff's face. We sat in the entrance, letting our eyes become accustomed to the half-light seeping down from outside, and gradually became aware of a steep slope dropping off into the mountain and spreading out into a great room of blackness. Strangely fantastic stalagmites slowly took form in the darkness, clustered about on the steep slope, their somber grey colors merging softly with the void below. In the quiet we could hear the steady drip, drip of water where somewhere on the slope another stalagmite slowly built up to join its fellows.

"Looks kinda spooky."

"Spooky or not, let's see what we've got."

The steep slope down into the cave from the entrance was covered with loose, broken rocks that fell ages ago when the final subterranean cave-in opened the natural entrance. And it was treacherous. Dripping water made many spots as slippery as an otter's mud-slide down a river bank.

"Watch that spot there," called back Tommie, who was leading. "It's slick as . . ." and then his feet went out from under him, and he tobogganed a dozen feet down the slope on back and shoulders, legs in air and arms flailing, trying desperately to protect the lantern from disaster. He climbed ruefully to his feet, the lantern undamaged, his cap lost somewhere in the darkness.

"Damn it!" he exclaimed. "See what I mean?"

I picked up his cap and we skittered to the bottom of the slope. Our eyes had slowly become accustomed to the gloom, and we could see that the cave here was a large auditorium, with two corridors leading off into the darkness. One went sharply to our left; the other angled slightly ahead to our right.

We turned into the left corridor, which sloped rather steeply downward for several score yards before leveling off into a large circular room whose farthest dimensions faded out beyond the glow of our lights. The ceiling was only a murky detail probably fifty feet overhead. To our right, close against the wall beneath a lower hanging section of the roof, stood a giant stalagmite which joined with the ceiling to form a massive column six times our height. A little farther beyond a great mass of brown flowstone covered the wall from ceiling to floor, like a frozen cascade of brown caramel.

"Boy, oh boy!" Charlie exclaimed. "Our Sunday school taffy pulls never came up with anything prettier than that."

Smaller stalagmites dotted the floor, which sloped gradually downward into darkness. As we advanced, the walls closed in gradually, until about 300 feet from the entrance of this corridor we apparently came to the end. A great hillock of flowstone towered above us, built up tier upon tier like a terraced Chinese mountain and capped with a group of stalagmites which seemed like the towers of an ancient castle.

"And Mohammed came to the mountain," I heard Tommie mutter. "Or we come to the end of the way."

But we discovered a smaller corridor leading off to the right. It seemed to promise great things, but like some promises, it just did not develop, pinching down to an end in about fifty yards. Flowstone covered its walls, and stalagmites and pillars studded its length. All were old and dry, scaling off to the touch.

We returned to the hillock of flowstone, and after clambering about halfway up its side found a way around it into a stalagmite-filled corridor which extended for probably 300 yards before ending in a blank wall. Close up against a side wall we found where someone had excavated a deep hole.

"I'll bet this is where Bill Burnet found his fossil bones," I remarked.

"What did he say they were? A caribou?"

"Sure was a long-lost caribou to get this far from the north pole," Tommie snorted.

"A caribou in a cave!"

"Probably one of those saber-toothed tigers I've heard about dragged him in here."

As we started back from this left section of the cavern, Tommie stopped before some of the great masses of flowstone covering the wall. "This sure is an old cave," he remarked. "It looks like it has been dead for centuries. Look here. Water erosion has cut these big channels in the flowstone, and I'll bet water hasn't run down these walls in years."

We made our way back to the main auditorium, with the ray of blue light filtering down through the entrance, and turned into the other corridor angling off to the right. It was about fifty feet in diameter, with a few small stalagmites here and there on the floor. Our lights revealed a scattering of stalactites along the ceiling. After about forty yards a smaller corridor branched to the left, dropping at a slight angle. We followed it for almost 800 feet before it ended abruptly; like a round tunnel some thirty feet in diameter, it went straight as a string. The rocks underfoot were rough and dark grey; only a few straggly formations decorated its floor and ceiling, but flowstone of a dark brown hue plated the walls. We called it the Mole's Run.

"This cave is filled with promising dead ends," Charlie complained as we started back for the main corridor. Shortly after reaching that section we approached a large stalagmite which dominated the scene. Beginning as a gently sloping mound of dark-brown onyx, it reared up for a dozen feet, then branched into two large pointed knobs. As we stood there admiring it, Charlie scrambled around behind it, his lantern throwing its bulk into strong silhouette.

"If that's not a rhinoceros sticking his head up out of the mud, then my name's not Trader Horn," Tommie said.

Darkness beckoned from a large corridor behind the Rhino. We had to make our way down into and out of a deep depression to gain its entrance; it seemed to run in a direction parallel to the Mole's Run, but in the opposite direction. Seeming to guard its passage, a great column stood just to the right; its circumference had become so great that for almost its entire height it had grown against and become solidly attached to the wall behind. Fully twelve feet in diameter and at least thirty feet high, its sides were decorated with beautiful deep flutings. Whites and tans and soft browns blended along its surface to make a thing of startling beauty.

"How about the Sentinel for this one?" Tommie asked. "It seems to be guarding the way."

Passing the Sentinel, we continued into the corridor, the floor of which sloped gradually upward. Within a short distance our lights picked up a ghostly white stalagmite ahead in the gloom, almost filling the passage. Once it had stood proudly erect like a tapered oriental temple, its sides built up in graceful fluted terraces. But at some time in the distant past its growing beauty had been its downfall, for the ground at one side had collapsed beneath its weight, tipping it dangerously out into the corridor. Tiny terrace after tiny terrace of flowstone had built up about its base, once again firmly cementing it to its foundation.

152

"And now, ladies and gentlemen," I joked, "I give you the Leaning Tower of Pisa."

Standing just behind the Leaning Tower, seemingly blocking the passage, was our most unusual formation of the day. Beginning at the ceiling as a heavy, broad mass of flowstone, it bulged out and down from the wall for a dozen feet, forming a great circular concave rim, or lip, like that of a monster toadstool. Back up under the lip, at the apex of the cone, the flowstone began again, falling to the floor in a series of gentle, ever-widening cascades, to form the toadstool's stem.

153

154

"Boy! What a hunk of steak could be smothered in that mushroom," Charlie said.

We discovered a small passage around the Toadstool which led into a beautifully decorated room some sixty yards long and half as wide. Tiny, delicately shaded brown stalactites decorated the ceiling, and great masses of colorful flowstone covered the walls. Stalagmites stood about in profusion. Coming into the room we edged around a great column of white limestone whose great bulk joined the wall from floor to ceiling. A few yards in front of it a thin partition of white serrated flowstone extended out from the wall for several feet like a jutting marble slab. Delicate shades of browns and tans and creams mingled with pure white, making this a room of fantastic beauty.

155

"Well," I said, playing my light about. "The toad's gotta have a place to live. This must be his palace."

"A toad's palace," Tommie snorted. "And I suppose all he needs is the beautiful maiden's kiss to turn him back into the handsome prince!"

The corridor ended with this room, and we made our way back

to the main passage and the Rhino. A dark opening beckoned from the shadows across the way. We walked along it for about two hundred yards, finding it uniformly about fifteen feet in diameter and barren of decorations. Suddenly it made a right-angle turn to the right through a small opening, and we stepped into another passage somewhat larger than the one we had just left, and parallel to it. Both seemed to parallel the Mole's Run we had left behind.

The corridor we found ourselves in was the scene of an ancient subterranean disaster. Great mounds of jagged grey rocks littered the floor, stacked in many places nearly to the ceiling maybe thirty feet above. Many of the rocks were loose, making exploration hazardous. Tommie poised atop a large boulder, preparatory to leaping across to another.

"Watch it, Tom . . .," I started to shout, but before I could finish, the rock he was standing on turned beneath him, and he sprawled in an unfinished jump, barely grabbing the rock ahead with one arm, his other holding the undamaged lantern triumphantly aloft. The clatter of his fall echoed down the corridor.

"You hurt, Tom?" Charlie asked.

"Nothing but my dignity," he answered dryly.

The passage extended only about sixty yards, black and gloomy, so we headed back along it toward the main cave. There were no beautiful formations here, only the somber, broken, grey rocks. We came back into the main corridor about fifty yards beyond where we had left it.

Directly in front of us, just looming out of the shadows beyond our lights, stood a beautiful column spanning the distance from floor to ceiling. We picked our way across an ancient, powdery bed of bat guano to examine it more closely.

"Surely," I thought, "here is the wonder-working of underground water at its best."

Gracefully, yet majestically, this mighty stone pillar swept up to the ceiling, its sides a maze of symmetrical flutings like folds of heavy draped curtains. Beautiful browns and tans and oranges blended softly with the pure white. Water from above trickled over its surface, and it shone and sparkled under our lights.

156

"Look how it seems to support the ceiling," Charlie said. "The Pillar of Hercules couldn't do a better job."

"This must be the big formation Bill said he found the ancient fire pit near, because there's where someone's been digging," I pointed out. "Sure would like to find some pieces of that Pueblo Three pottery he claims he found here."

Beyond the Pillar of Hercules the main corridor swung slightly to the right. We found the floor covered with a deep deposit of ancient guano, coated in many places with a crust of flowstone of varying thickness. Often we broke through ankle-deep into the dry guano beneath. Hundreds of embryonic stalagmites dotted the surface of this crust, ranging from small knobs to budding formations a foot or more in height. All were growing, their wet sides resembling highly polished marble under our lights. Most of them revealed tracings of delicate yellows and tans, and were translucent to a light held behind them.

The Pillar of Hercules had hardly disappeared into the shadows behind when another massive group of stalagmites emerged from the gloom ahead, stretching for about thirty feet to shelter the entrance of a small cove-like chamber which held many beautiful limestone decorations. Almost the entire row of guardian stalagmites were a dark brown, but the extreme right group gleamed white as snow, as though someone had poured thick cream over them which had frozen into beautiful rippling cascades. We called the group the Guardian Towers.

Out in the center of the corridor rose a steep mound of broken rocks and powdered guano. We clambered up its side to the pinnacle over forty feet above. Ahead of us loomed only a vast darkness, broken here and there by the faint, ghostly image of some stalagmite weakly throwing back the gleam of our lanterns.

"Man!" Tommie shouted. "We've got some cave here."

To our left a large passage dropped down at a steep angle into a great shadowy pit. Loose dirt and gravel covered the slope, and here and there patches of wetness from dripping water made the slope even more treacherous. It was rough going, and an occasional thunder of sliding rocks and a muffled curse proclaimed someone was having difficulty. Suddenly a rock turned under Tommie, his feet headed for the ceiling, and he was off on a toboggan slide down the rock slope, amid flying gravel and a miniature dust cloud. His lantern went sailing through the air to crash and extinguish amid the rocks. Charlie and I balanced helplessly and watched him come to a halt with a crash thirty feet below us. We were afraid to ask how he was.

He clambered slowly to his feet and glanced ruefully in our direction. "This is getting damned monotonous," he complained. That was his only comment. I picked up his lantern and we clambered down to join him. The seat of his trousers and most of his shirt were in ruins, along with various sections of skin they were to have covered. Fortunately we carried extra mantles for our lanterns, and after installing new ones on his bent but intact lantern, we found it burned at almost peak efficiency.

We reached the bottom of the pit, far lower than the main cavern corridor, and found the passage continued into darkness at a slightly upward slant. Great boulders cluttered the floor, and the going was not easy. The walls of the passage became narrower and the ceiling lower; soon its diameter was not more than a dozen feet.

The overall color changed from brown to a dull greyish black. Smooth flowstone covered the floor, but as we advanced this smoothness became broken by tiny, thin, upraised walls, or dams, about an inch high, winding here and there. They grew larger as we continued, finally standing about eight inches high, like thin, scalloped dikes of stone. Then we came upon another of the stranger scenes in this fantastic cave.

162 We seemed to be gazing down from a great height across a dense forest of black spruce. As thickly as they could cluster and still retain their individuality stood hundreds of rough-surfaced, tiny stalagmites about four to six inches high. From a round, broad base each tapered gracefully to a fine point, exactly resembling an evergreen. They were a dull smoky black.

"The Black Forest!" I exclaimed. No name could be more fitting.

Just beyond the Black Forest stood a series of five beautiful dikes over a foot high, and in between each clustered many of the tiny stone 163 trees. They were a fitting climax to this side excursion from the main cavern, for this corridor ended here.

The clamber out of the Great Pit exhausted us, so we stopped at the base of Guano Mountain to eat our lunch and take a much-needed break. I heard a snort from Charlie, and knew he had dropped off to sleep.

"Well," Tommie exclaimed, rising wearily to his feet. "We came to see this cave, so let's get with it."

We circled Guano Mountain and descended into the vastness

beyond. Here truly was a Great Room. We picked our way in and out 164 among thickly clustered stalagmites, many of them towering above our heads. The walls dimly visible in our lights were a great conglomeration of beautiful fluted flowstone. High overhead the ceiling was indistinct, but we caught the gleams of many white stalactites hanging there.

We had gone about four hundred feet into this great chamber when out of the gloom ahead loomed a mighty, graceful stalagmite. 165 Tier upon tier of almost white limestone it towered up, each tier a beautiful section of deeply grooved, delicately paneled stone. Its double-spired peak was almost lost in the gleam above our lights, but we could see that it ended several feet beneath the ceiling, which was seventy or eighty feet overhead.

"Carlsbad Caverns never had a totem pole like this one," Charlie exclaimed. So we called it the Giant Totem.

Directly to the left, over against the wall, a striking formation stood out. On either side heavy frozen waterfalls of dark-brown flowstone covered the wall, yet this individual stood out in pure white 166 splendor between them. Its sides were as smooth as polished porcelain, its top a dome of gleaming white. Two black holes in its side, one above the other, gave the appearance of windows.

"The White Pagoda, maybe," I suggested. Hanging many feet above its dome was an oval, mushroom-like canopy with deeply serrated underside.

While Tommie and I were examining the Pagoda, Charlie had wandered off beyond the Giant Totem where the floor of the cave had leveled off as smooth as a ballroom.

"Hey, fellows!" he called. "Come here. I've found the Chinese Wall."

The resemblance was there. Standing about a foot high at its tall- 167, 168 est portions, it was higher than the Chinese Wall in Hidden Cave, and was built up of a series of small, reversed loops, like the ribbon candy so dear to the hearts of children. It wandered indiscriminately, like a leisurely snake, across the level floor, encompassing at least one thousand square feet. In many places it wound back upon itself, forming small corrals. The floor was almost black, covered with thousands of tiny knobby growths about half an inch high.

We followed the Chinese Wall to the left to a point where it doubled back just in front of a corridor leading past a white, squat stalagmite. Entering this passage, we found ourselves in a room that left us speechless. Here, if ever, Mother Nature loosed all bonds in order to create for herself a fairyland of her own.

The walls were buried behind a maze of colorful flowstone waterfalls, great hanging masses of stone draperies, delicately tapered stalactites, mounds and protuberances and niches. Color was everywhere —blacks, dark greys, white, browns and olive green and orange, even delicate shades of red. Broad-based, round-topped stalagmites stood about, covered with a mantle of smooth creamy flowstone like the icing on cake. One had a long tapered spike growing from its rounded top, like a gigantic helmet of the old German army. It must have been at least eighteen feet tall.

"I guess we finally got to fairyland," Tommie breathed softly.

Slowly we advanced into the room, trying to absorb its beauties. Then to our right we suddenly beheld this fantastic room's crowning achievement. A great stalagmite stood there, dominating the entire scene. No one could mistake the resemblance—like a giant evergreen it seemed, covered over with a thick mantle of white.

"The Christmas Tree!" I exclaimed.

Its under portions ranged from a creamy tan to a dull olive green. At intervals overhanging edges exactly resembling snow-burdened branches peeked from beneath their covering. And how like snow that mantle looked, for millions of tiny calcite crystals over its surface caught the rays of our lanterns and threw them back like myriads of diamonds.

We measured it and found it to be slightly more than eighteen feet in diameter at the base, and fairly accurate estimation placed its height at forty feet. The floor here was the site of an ancient underground lake, and the receding waters had left its surface soft and spongy to the eye, but hard and rough to the touch. Remnants of the lake still lingered in the deeper section, and we found the water ice cold, palatable, and most welcome.

Across the room from the Christmas Tree, hanging in an alcove all its own, was a formation which at first glance seemed to defy the laws of gravity. Like a great stone tear it hung, apparently supported only by its connecting stem at the ceiling, its base several inches above a round knob on the floor rising to meet it.

"Hey!" I exclaimed. "Remember that woman in our mythology book who wouldn't stop mourning the loss of her children, and the pitying gods finally turned to stone. This must be one of her tear drops." So we called it Niobe's Tear. Closer inspection showed that it was not actually hanging free, but was connected for part of its length near the top to the wall behind. It was fully eight feet long and at least six feet in diameter at its greatest breadth.

171

We were loath to leave this beautiful room, but unexplored portions in the darkness beyond tempted us. Tommie was leading the way out, and as he turned into the large passage that led into the main portion of the cave, he stopped suddenly.

"Fellows, I believe we're haunted," he said.

A monster crouched there in the entrance of the passage, seeming to dare us to pass. This was the other side of the squat white stalagmite we had passed upon entering the room, but what a different aspect it presented now. It was an old formation of dirty brownish hue, wide and flaring at the bottom and tapering quickly to the top. Over it lay a mantle of white flowstone like a shrouded hood. It took no strong imagination to see an evil, bestial face leering out from beneath the hood, the broad, ugly mouth filled with yellow, straggly teeth, the flattened, wide-orificed nose.

172

"Looks like something out of a bad dream," Charlie said. We remembered that Burnet had spoken of a formation in the cave which he called the Clansman, and we knew this must be it.

Back in the Great Room we skirted the far side of the Chinese Wall and a large narrow corridor opened up before us. The entrance was cluttered with great boulders, and we scrambled over these, narrowly avoiding a deep chasm to our left as we entered a large circular room. The narrow pathway leveled out, and a slender white stalagmite no larger around than my leg but fully fifteen feet high stood before us. Passing around this, we circled a large jutting shoulder protruding into the room. A great shape loomed up before us, but even when our lights dimly outlined it we still could not believe what we saw. Something that huge simply could not have been built up by dripping water. We felt that we had certainly found the world's largest stalagmite.

Its base was so massive that we first thought it a part of the wall. With heads thrown far back we tried almost vainly to see its top, for the glow of our lanterns only faintly reached that height. But the spot of light from an electric torch picked out the tapered peak high up against the ceiling.

"If I didn't see it I wouldn't believe it!" said Charlie—the old cliché suddenly ringing true.

Fluted section upon fluted section its massive brown shape reared up into the darkness. Twenty or more feet in diameter at the base, it swelled progressively for another twenty feet, then tapered gradually to a rounded point just connected with the ceiling. With the aid of a rangefinder on one of our cameras, we measured the distance from the base to a spot on the ceiling. The twenty-five-meter point did not quite bring it into focus. Eighty feet! No wonder we could hardly believe our eyes! It was at least five feet higher than the stalagmite in Cottonwood Cave. And what was the Giant Dome in Carlsbad Caverns, with its sixteen-foot diameter and sixty-foot height, when compared with this colossus. That was the name I thought we ought to give it. Draped around its base were great folds of onyx, deep enough for a man to step completely within. A light behind a fold revealed beautiful transparent shades of rich browns and deep reds.

Beside the Colossus stood a formation that would have been massive in other surroundings but was dwarfed here by comparison. Its sides were deeply grooved and capped with a perfect white dome like the Capitol in Washington. Grouped about on little ledges at various heights on the wall behind the giant were many small, dark-grey stalagmites, each capped with a layer of white limestone like a dripping ice-cream cone.

"Fellows, I've got to have a picture of this," I wailed, "but I don't think I've enough powder left."

I rolled up my paper fuse and poured out the remaining flash powder. "That won't do it," I cried.

"I have some left in my bottle," Tommie offered.

I poured out some of his, looked at the pile and then at the huge thing towering before us, and upended the bottle.

"Not all of that!" Tommie exclaimed. "You want things to start parting company?"

I shrugged and lighted the backside of the paper. The explosion of the powder boomed hollowly in the great room, and the instant flash captured a scene of indescribable beauty. And no stalactites came crashing down.

"That ought to do it," I said, folding up my camera.

"Fellows," Charlie said, "you know how long we've been in here? I think we ought to be heading out."

Suddenly we realized how tired we were. It was a long, rough scramble to the outside world, but eventually a blue glow far ahead betokened the end of our explorations, for it was daylight we saw seeping down through the entrance. As we paused at the foot of the steep slope leading up to the opening, the many fantastically shaped stalagmites clustered about seemed eerie and unreal in silhouette against the soft blue light; they looked like monsters from another world lost in a deep fog at twilight. Finally we stood once again in the hot, dry air of Slaughter Canyon, our eyes squinting in the brightness.

174

"Well," Charlie said as we started down the mountain to our car, tiny in the distance below. "It's no Carlsbad Caverns. But it sure as hell is no 'dirty little bat hole'."

As we hunkered down around the fire drinking coffee, furnished as usual by Tommie, Charlie suddenly said, "It's such a grand cave, and I think it deserves a better name than simply New Cave, which Bill Burnett continues to call it. New Cave just doesn't describe anything."

"That's funny," I put in. "I've been thinking the same thing. You know, we've been in a lot of caves in these old Guadalupes, and not a one of them is named after the mountains. I think Guadalupe Cave would be a fitting name for the cave, and a great tribute to these old rocky hills."

"That name sure 'nuff hits the spot," Tommie said. "Guadalupe Cave she'll be, if we can just swing the deal." Thereafter, we always referred to it as Guadalupe Cave, in an effort to abolish the lack-lustre name of "New."

That was only the first of several fascinating trips I made into this great cave. Just a week later, as soon as we got off work, Tommie, Charlie, Sonnie, and I headed for the cave. The last rays of the sun burnished the high precipices along the rim of Slaughter Canyon across the way as we stopped in the entrance for breath. The great cliff we called the Kneeling Elephant stood out in sharp relief. Our chief goal on this trip was photography, but we also did some exploring, finding several areas we had missed the week before.

We had made it to the Great Room—Sonnie said we should call it the Ballroom, the floor was so level—and were walking along the flowstone-covered wall near the White Pagoda. As we stopped to admire its smooth, shining beauty, Charlie noticed a small opening leading around behind it that we had completely overlooked on our previous trip.

"Looks interesting," Tommie remarked. "Think we can get up over that slick front into it?" It was a wild scramble, but we made it and emerged into a narrow, high-ceilinged room of rare beauty. Slender, beautifully tiered stalagmites grew in profusion, and the ceiling was studded with delicate stalactites of many hues. No section of the rock wall was visible, for all was covered with heavy masses of gently undulating flowstone ranging from creamy white to a deep rich brown.

Almost in the center of the room, hanging suspended over a group of slender white stalagmites, was a thin sheet of snowy onyx at least fifteen feet long and perhaps three wide, and never more than half an inch thick. Like a heavy, curved-edge blade it hung there.

"Just like a guillotine," Sonnie muttered. "A guillotine in a white palace."

We continued along the passage, which shortly made a sharp bend to the left, and we passed by several large mounds of bat guano, wet and slippery from dripping water. Here the dark browns predominated, especially in the masses of flowstone, some of which hung in sheets resembling heavy velvet draperies.

Again the passage turned sharply to the left, and several feet beyond we stepped out into a great auditorium.

"Something sure familiar here," Charlie exclaimed. "Look there!" Ahead of us in the dim lantern light loomed a familiar shape. It was the Great Totem. We had made a circle and come back into the Great Room some sixty yards from where we had left it.

We turned back up the wall toward the Christmas Tree Room, and within a few yards a large cleft beckoned us. It proved to be a corridor about twenty feet in diameter which extended back for only about fifty feet. At one side of the room a large pool of crystal pure water invited us to quench our thirst. At the back of the room was a scene of catastrophe. Broken stalagmites and stalactites lay tumbled about in great confusion, stacked upon each other and shattered. But this had happened long ago, for now all were firmly welded together by the flowstone and covered with a creamy layer of wet, glistening frosting, as appetizing looking as the caramel center of a candy bar.

177

"I think we should have a picture of this wreckage," I said, unholstering my camera.

We spent the entire night photographing. When the smoke from the flash powder made picture taking impossible in one area, we merely moved to a new one. When the last of the powder had gone up in smoke, we made our tired way to the entrance. After enough time, we knew we should have been there.

"Where in the hell is that opening?" Tommie asked in exasperation. "We should have been there long before now."

"We couldn't have missed it," I replied. "There's only this one way out."

"You know," Charlie put in. "I thought something looked familiar back there a ways. Isn't this the way into the Leaning Tower room?" He was right, for a short distance farther our lights picked up that stalagmite leaning out into the corridor.

"You know what's wrong?" I asked. "It was daylight the last time we came out, and we could see the light outside. It's still dark out there, and we missed the entrance."

Feeling rather foolish, we backtracked until we recognized the slick slope leading up to the entrance. As we stepped out into the cool night air, the first grey hints of dawn glimmered over the Staked Plains to the east. We were bushed by the time we got down to the car, and Tommie yelled, "Coffee! I gotta have coffee!"

We got a fire going and he brewed up a big pot of his specialty. No one made coffee over an open fire like Tommie Futch. I think it was all that sustained us into town.

In the ensuing three weeks we talked so much about "our" new cave and its beauties that the women folk began pestering us to take them to see it. So on October 16, 1938, four couples of us were plodding up the trail to the entrance just as the sun peeked over the eastern horizon. Our party consisted of Tommie Futch and his wife, Mary; Charlie Stevens and his wife, Claire; Douglas Evans and his wife, Mae; and my mother, Zena, and me. Although in her mid-fifties,

Mother was wiry and agile and knew she could make the trip. And she was right. In fact, she was standing in the entrance when some of the others arrived.

We did not try to find any unexplored sections this trip, but concentrated on showing the ladies what we had been bragging about. However, we did stumble on something new just beyond the Christmas Tree Room where the Chinese Wall swung closest to the side of the room. A small alcove opened there, and in ages past a great stalagmite had toppled over at the entranceway and now lay like a fallen log bridging a round depression beneath it. The deep flutings which had once been vertical decorations down its length now lay parallel to the floor. We clambered down into the depression beneath and looked up at its underside, like a bridge above us. The vast age of the past disaster was attested to by the great mass of white flowstone that had formed from a recess above and now covered the top side of the prostrate stalagmite to a depth of a dozen feet, oozing down its side in places to form small stalactites from its under side.

"Like a fallen giant, buried and forgotten," Mae said.

We clambered up over it to a small dark opening behind, and found ourselves in a series of tiny rooms containing beautiful sections of white flowstone draperies. Here everything glistened wetly, and every little depression held water.

The ladies were as entranced by what we showed them as we were. Even Mae, who was beginning to be visibly pregnant, forgot her condition and joined enthusiastically into the activities. Mother was most impressed by the giant stalagmite—the Colossus—which we saved for the last to show them, and insisted she had to have a picture at its foot. So once again I poured all the remaining flash powder out on my paper fuse, and once again the great auditorium boomed and reverberated when the charge went off, and once again nothing came loose.

But the girls were weary when we finally got down to the cars in the canyon, and Tommie brewed up another of his famous batches of coffee. We had just settled down to enjoy it when a pickup truck came tearing up the canyon and slid to a stop beside our cars. A tall, somber cowboy got out. We had no difficulty seeing the big revolver holstered at his hip, nor the 30-30 Winchester nestled in a rack behind the seat.

"Hi," Tommie greeted, friendly like.

"What do you folks think you're doing here?" the stranger asked, by way of greeting. About then I recognized him as Johnny Colwell, who owned the ranch down at the mouth of the canyon. He seemed a little taken aback by the sight of ladies in our party.

"Oh, we were just up looking at the new cave," Mary laughed.

"Don't you know you aren't allowed in here without permission?"

"I don't guess we knew that," someone replied.

"Well, it's true! And you're going to have to pay me a dollar a head for going into the cave."

Tommie laughed and poured an extra cup of coffee. "I don't reckon there's a whole dollar among the lot of us," he drawled. "Here, have a cup of coffee instead."

I guess the rancher realized we probably knew he didn't have the rights he claimed; so he took the coffee and haunched down amid us to drink it. I heard Mother breathe a sigh of relief.

It was late in November—the 27th, in fact—before we got back to the cave. Again it was one of our all-night trips, and the sun had dropped below the rim of the canyon in the west by the time we reached the cave's entrance. As my caving companions this trip I had Tommie Futch, Charlie Stevens, Sonnie Kindel, J.D. "Static" Burke, Wayne Stell, and Doug Evans and a dude friend of his from the East, whose name I never did get straight. Photography was our chief objective, but of course we would not turn down a little side exploring if the chance offered.

After a side trip into the Leaning Tower section, we skirted the right-hand wall of the cave, a portion we had tended to neglect. We passed the great Towers guarding the pretty alcove behind and started up the great guano slope with its thin crust of flowstone. Doug stepped off to one side to admire some small orange-colored stalagmites, and we heard a cracking noise and a loud grunt. He had broken through and dropped to his waist in a cavity in the dust-dry guano. Frantically trying to get out of the hole, he pawed at the thin crust, which broke under his efforts. I could not help thinking of an ice skater down in a hole in thin ice which broke in front of him with his every effort to clamber out. The powdery guano filled in around him, and in no time, it seemed, he was covered to his chin in the stuff, with only his eyes shining in our lights.

"Take it easy, Doug," Tommie shouted, as he got down on hands and knees and crawled up with extended hand. Doug reached it, and Tommie hauled him out, shaken and guano covered.

"Christ!" he exclaimed. "I didn't think that stuff had any bottom."

"Well, I can tell you one thing," Sonnie snorted—he did not bear too much love for Doug in the first place—"I can sure say I've seen you with B.S. up to your neck, and I don't mean the kind that comes from bulls." Relations between them were strained for the rest of the night.

We passed the largest mass of almost coal black flowstone I had ever seen, covering the wall from floor to ceiling, and just beyond a rather large alcove opened into the wall. It was round, almost chapel-like, perhaps twenty feet in diameter, and in the center was a black hole about twelve feet across. It had once been a pool of water, for onyx had formed in great scalloped sheets along its entire circumference, like great lily pads of stone, and as the water had receded, it had left them hanging in space, solidly cemented at their back edges to the brink of the pool. In many places they overhung as much as five feet.

With the curiosity that killed the cat, I decided I had to see the bottom of the hole. Getting down on all fours, I crawled out to the edge of one of the great lily pads. It seemed solid enough. I reached the edge and, holding my lantern over, leaned out to peer down. My 200 pounds were too much. Before I had a chance to see bottom the lily pad snapped off at the back edge, and down I went.

I do not know how many things I thought of going down. I remember the sides of the hole around me like a cistern, the way lightning lights up everything for just a second, then is gone; that is the way the walls flashed at me in the lantern light, and they looked rough and hard. I did not know how deep the damned hole was—it might be a fathomless pool of water—and I was headed head first for the bottom.

"Don't fall on the lantern," I remember thinking.

Then I saw bottom coming up; there was a small hole right below me. I jammed the lantern into it, and crashed, amid a great clatter as the lily pad I rode down shattered to bits on the rocky floor beneath me.

"Jeez, Sam!" I heard Tommie's voice above. "You all right?"

I was positive I could not get up, and astonished when I did. The pits of hell could not have been darker, for the lantern was a complete washout in the hole where I had crammed it. Then light flickered down, and I saw white faces at the edge where my lily pad had broken off. I was in for another surprise, because they were not more than ten feet above me; I could have sworn I fell fifty.

"You hurt?" Tommie called again. I took stock and decided I was not. As a matter of fact, I hardly had a scratch. I could not believe it, and neither could they.

"From the racket you made when you hit," Wayne said, dryly, "I expected to sadly shovel up the remains and start a search for survivors." I did have a thumb knocked out of joint when I rammed the lantern into the hole, and some scorched skin across the top of my hand. The lantern was a total loss; I did not even bring it out of the hole with me when they helped me out.

Just beyond the pit of my downfall a large corridor branched off at right angles to the Great Room, which we were just entering. It extended for about 200 feet straight ahead, but proved to hold little of interest. About 50 feet beyond the entrance of this one a smaller corridor branched off, parallel to the last one, but it also had little to interest us.

Still circling the right wall of the Great Room, we came into sight of the right loops of the Chinese Wall. Off to the left, in the center of the room, the Great Totem loomed ghostly in our lights. Here another corridor right-angled straight ahead. We followed it for about 500 uninteresting feet before it dead-ended in a blank wall.

Back in the Great Room, we made a side trip into the room of the Colossus; its majestic splendor, like a monarch of all it surveyed, never failed to awe us.

While showing the others the Fallen Giant, Tommie discovered a small passage leading down just beyond and beneath it. We crawled down through the opening into a triangular shaped room, the greater portion of which had once been a large underground pool. The old waterline still showed plainly, encircling the room at a height of about eight feet. Now, however, only a small circle of water remained in the deepest section.

179

"Water!" Tommie exclaimed. "Boy, am I ready for that." We were thirsty, for we had found no water in our circle of the right hand wall.

From the pool of water both floor and ceiling of the room sloped up at a sharp angle. With some difficulty we crawled up the slick surface into a larger room with a level floor. Up at the far end, at the head of another steep incline, appeared a tiny opening; we clambered through that, finding ourselves in a narrow passage through which we could just squeeze our way.

"Talk about Fat Man's Misery!" Static exclaimed, reminded of a passage Jim White had named in Carlsbad Caverns years before.

The passage widened slightly, and we realized we were inching our way around a huge stalagmite which almost touched the wall. Finally we stepped out from the narrow confines into a large room with a spongy looking, cream-colored floor. We turned to look back at the big stalagmite we had just come from behind.

"Well, I'll be damned!" Sonnie exclaimed. "The Christmas Tree."

We sat on the edge of the pool and finished our sandwiches, washed down with the cold water at our feet. I noticed a small opening high up in the wall above the Christmas Tree. It seemed to invite looking into. The wall was rough here, studded with several projections and small stalagmites growing on little ledges, all giving promise of potential handholds.

"I believe I can make it up there," I said. Most of the others looked dubious.

"Maybe so," Tommie answered. "Here, take my lantern and have a whirl at it."

I got rid of my camera and other bulky equipment, grabbed the lantern and started up the wall. About the first twenty feet proved to be no problem, with plenty of hand and foot holds; then I found myself more or less stranded and forced to plot my further course. I was standing on the base of a short, thick stalagmite that jutted out from a small ledge for about two feet, then angled upward, making a sort of hook. From where I stood I could just reach another short stalagmite on a ledge above. With the lantern in my left hand, I grabbed the stalagmite above with my right hand and started pulling myself up. I was just about half way between where I had been and where I was going when my handhold snapped off.

I tried to whirl around so I would fall facing away from the wall, but my foot caught for an instant on some projection, throwing me sideways, and I fell belly straddle across the hook-like stalagmite I had been standing on, hanging there like a sack of meal across a mule. But

I did not hang there long. I could smell cloth and skin burning and I did not need a lawyer to tell me it was mine. I had fallen squarely on top the lantern.

I let out a squall like a branded stallion and heaved off my landing place, slithering down the rough wall to land at the others' feet. In the process I left a patch or two of hide along the way and the better half of an outraged shirt. The offending lantern almost beat me down.

"I guess that wasn't the way, after all," Tommie muttered as he applied first aid to my various cuts and abrasions and a burn across my stomach that was rapidly rising to a blister. Then he picked up his lantern—what was left of it—shook his head, and remarked dryly, "This has been a tough day on lanterns. Scratch number two."

"It just hasn't been my day at all," I wailed. "What a mess."

I had had enough for one trip, and the others seemed to feel inclined the same way, so we headed for the entrance. This time we did not miss the slope up to it, because daylight was filtering in. We had spent the entire night in the cave. The sun was already up when we stepped into the outside world, and Tommie's famous pot of coffee over the open fire revived our spirits.

It was almost five months before I went again to the cave, on April 23, 1939. My old mountain-climbing friend, E. F. Brunneman, wanted to see a good cave in its wild state, and I could think of none better than this one. I also wanted to go back to make a rough map of the cave, so I carried a clip board, drawing papers, and a compass, and we spent the day stepping off the distances and marking directions. I knew it would not be the most accurate map in the world, but it would give me a working plan of the cave's general layout. I was satisfied with it, and Brunne was happy with his first experiences at wild caving.

On the way out, we were about half way up the slippery slope to the entrance, in full light of day, when I noticed a small corridor branching off to the left that we had overlooked on previous trips. It was about fifteen feet in diameter, and as we made our way along, it swung steadily to the right. After about three hundred feet, I thought I noticed a soft glow ahead. I stopped and turned off my flashlight.

"That is daylight!" I exclaimed. Sure enough, ahead was another entrance to outside, a small hole through which a man could probably squeeze. It was at the top of a straight drop of a dozen or so feet, and we could not get to it without a rope. So we retraced our way and left the cave by the usual large entrance.

Over a year was to pass before I made another trip into the cave. The date was May 26, 1940. Much had transpired meanwhile: The country was well on its way out of the depression; I had married a Missouri girl from my college days; Europe was in turmoil. My college friend, Vincent Lockhart, had married, was now in the army, and on a furlough visiting us in Carlsbad with his wife, Helen, so we decided a day of wild caving was in order.

Things had changed at the cave, too; a steel bar barricade had been erected across the entrance, and it was securely locked. However, I had an ace in the hole. I had heard about the barrier, so carried a lasso. The four of us made our way around a shoulder of the mountain, and by a little careful plotting I was able to find the small entrance that Brunne and I had discovered from the inside a year ago. The other three got through easily and with aid of the rope reached the bottom of the drop inside; with some difficulty I managed to squeeze through the hole too, and we were on our way into the cave.

We found changes within also. A guano-mining enterprise had leased mining rights in the cave; they were the ones responsible for the gated entrance. Heavy steel cables stretched down the entrance slope and back into the cave. We found one firmly anchored about the Rhino, which was much the worse for wear. All the cave formations in the entrance area were covered with guano dust, and great excavations were all about in the ancient guano deposits. However, most of the mining activity had been concentrated in the front half of the cave, and we found the more beautiful sections in the back half and in the side passages relatively free of damage. There were some evidences of vandalism seen in the stubs of broken stalagmites, but not too many.

After enjoying the beauties of the Christmas Tree Room and eating our lunch beside the pool of water, which was still clear and good, I suggested a trip up to the high alcoves we called the Mezzanine. We had discovered this section on the November, 1938, trip, after a tough, steep scramble up a series of ledges in the Great Room near the entrance to the Christmas Tree Room. The girls took one look up the precarious way they would have to go.

180

"Not me!" Helen exclaimed. "I love exploring, but I'm no monkey."

"Me neither," my Elizabeth agreed.

"O.K." I answered. "I'll tell you what. You two take the lanterns and wander around here in the Great Room, and Vince and I will take the flashlights and go on up. It should be quite a sight up there, with you lighting up the area so we can see."

It was. The climb up really was not too bad, taken easily. From the alcoves and the High Ridge, as we called it, we looked down into the Great Room below us. Even the top of the Giant Totem was on about eye level with us. As the girls wandered about, their lanterns disclosed in some detail the magnificence of this fine auditorium. The Giant Totem, almost in the center, completely dominated the scene, with the Guardian Towers looming up in the far left distance. At the right, against the wall, the White Pagoda gleamed; we had come to call it the F.F.F. after an obscene remark someone had made about some of its frankly sexy appearing characteristics. We looked almost squarely down upon the Chinese Wall directly below us.

"Boy, this is something!" Vince exclaimed. I could tell by his voice that the cave bug had bit him. "I wish the girls could see this."

"Yeah, but just remember we wouldn't be seeing it if they were up here with us."

When we got back down and joined them, they had had enough, and we made our way out to the Keyhole, as we named the small entrance through which we had come. The girls discovered that coming in down the rope was much easier than climbing out up it, but after some straining and encouraging pushes from behind they made it.

More than two and a half years were to pass before I visited the cave again. The calendar said October 4, 1942. Europe and the Orient were aflame. An army airbase was located at Carlsbad. There was little time for anything but war-oriented activities. Elizabeth's brother, Robert Fleming, had joined us in Carlsbad and was working with me at the airbase. A weekend all our own faced us, and we decided to show him a wild cave. The floods that had swept out of the mountains in the spring of this year had wiped out the road to the trail leading to the cave; in fact, the floor of the canyon and a great area of the plains out beyond had been swept clean by the water and were now a great sea of white rocks and gravel. We had to walk almost two miles just to

181, 182

183

184

get to the old trail leading up to the cave. Guano mining had ceased, but the gate was still locked across the main entrance. And we found the Keyhole had been plugged with cement.

"Maybe we don't go caving," Elizabeth remarked.

We scrambled back around the shoulder of the mountain to the main entrance, and closer examination revealed that, with a little digging, we could slip under the gate. I found quite a bit more damage to the front half of the cave from the mining operation, but the colorful back sections were still beautiful and mostly intact. More evidences of vandalism were seen. Most of the white-capped stubby black stalagmites—we had originally called them ice cream cones, then the Senior Colored Choir—that dotted the ledges behind the Colossus were gone; some still lay in rubble at the foot of the wall. In many places only broken stubs revealed where colorful small stalagmites had stood. Gone also was "Big Richard," the two-foot-high stalagmite that had stood in the corridor leading to the Leaning Tower, the one that had been the butt of many jokes because of its perfect representation of a phallus. But the cave was still beautiful, and still fascinating.

I was setting up for a picture of The Clansman, when suddenly Bobbie said, "Listen! I hear talking."

Sure enough, a light glimmered away back toward the entrance, and soon we were joined by a civilian employee at the airbase, a Mr. Belche, and a Sgt. Rufus Wood, who had heard of the cave and had spent most of the day looking for it. I had them join us in the picture.

I felt a tinge of sadness as we left the cave. The destruction of its beauty troubled me. And with the world in such a mess outside, I wondered if I would ever again wander through its unnatural beauties.

POSTSCRIPT

Fate decreed that I would visit the cave again—almost 32 years later.

March 3, 1974 . . . This time I was a member of a regularly scheduled party under the direction of National Park Service guides. The cave was now officially named New Cave, a part of the Carlsbad Caverns National Park. Limited tours by prior registration were regularly conducted. The cave was maintained in a "natural state". Tours were by lantern and flashlight. A park service road led to the mouth of the canyon, and an easy trail led from there, following gentle slopes to the cave entrance; the walk was much longer, but much easier, than our old scramble up the mounain.

The gate at the entrance was still locked, but short flights of stairs, some handrails, and gently sloping trails replaced the old slip and slide descent down the slope to the floor of the cave. Our guided tour followed easy trails through the main cave corridor, to the Christmas Tree Room and the room of the Colossus, now officially named the Monarch. Most of the mess resulting from the mining operation had been cleaned up; one or two holes in the guano were protected for historical purposes, and we were cautioned against stepping on a few well-preserved tire imprints the miners' cars had left.

We did not see the corridor to the Leaning Tower and the Giant Toadstool. The guide mentioned vaguely a room somewhere that had a Black Forest but was inaccessible; I do not think he knew where it was, and when I mentioned the Leaning Tower of Pisa a look came on his face that seemed to mirror his thoughts, "This dude thinks he's in Europe."

Everything was very orderly and arranged and well-planned. But I kept hearing ghostly voices out of the past, laughing at a fellow's predicament, yelling at the wonder of a new discovery, small-talking the way we felt. I heard again the crash of a giant lily pad shattering beneath me at the end of a fall, smelled again the burning of cloth and skin. Now, as the flash bulbs popped about me, I sniffed again the old acrid smell of burnt flash powder.

As I stepped back into the bright sunlight, the thought crossed my mind, "Yes, dear old Guadalupe Cavern, you've finally come of age."

But in my heart, I wondered.

Epilogue

CAVE EXPLORATION and photography slowed down for me in the fall of 1934, when I enrolled at the University of Missouri, but I never missed a chance to go underground when vacation time came around. Marriage followed closely on the heels of graduation, and all the old gang likewise became involved in family life and the problems of making a living, and caving suffered. Some of the cavers moved away: Dave Wilson back to Louisiana, Seth McCollom to Oklahoma, Wayne Stell to parts unknown. Ted Fullerton broke a leg in a freak accident in his home, and caving for him became a memory. Death has taken many: my brother B. A., Tommie Futch, Sonnie Kindel, Julian Shattuck, Sonnie Hagler, Bill Bryan, Carl Livingston, Jay Leck, Lawrence and Forestina Matthews, E. F. Brunneman, Marge Dwyer, Bacel Scott, Bill Liddell, Jim White, Bill Burnet, my mother Zena, and most of her caving friends. How long and how rapidly the list has grown.

A younger generation of cavers, eager and daring, now explore the Guadalupes, and the register of the new caves they find grows larger month by month. Today, this fantastic mountain range ranks as one of the hottest caving spots in the country, with every weekend finding cavers on and in its far ridges. Caving has not lost its fascination for me, and I still manage to get in some trips. No strangers to me are the recently discovered mysteries of such caves as Cave of the Bell, Cave of the Pink Dragon, Christmas Tree Cave, Lost Hunter Cave, Blowing Cave, Robbers' Loot Cave, and Dam Cave, to name a few.

But now, in the summer of '78, although the lure of the underground is as strong for me as ever, I must bow to 67 years and a football knee that grows more obstreperous each season. The young fellows in the Guadalupe Grotto of the National Speleological Society, of which I am immediate past chairman, still include me in plans for their explorations. But I find I must accept only the easier trips; the strenuous ones and those with long vertical drops are out.

Yes, "The old order changeth, yielding place to new . . . . "

But the memories linger. And as long as breath remains, my senses will throb to the fascination of a dark hole in a mountainside, the sound of dripping water in a darkness and a quiet so intense that each drop rings like a gong, the siren song of a virgin passageway, with a gleam of white stalagmites just appearing in the furthermost beam of a flashlight where the tunnel bends out of sight into the mysterious unknown.

That thrill will never die.

Carlsbad, Caves, and a Camera

Photographic Scrapbook

The photographs reproduced here represent a cross section of photographs made by the author, together with a few made by other photographers that are now in his collection. They date from the late 1880's through the first four decades of the present century.

Following the caption to each photograph is the number of the page where the narrative appears explaining what the picture portrays or why it came to be taken. The photograph number appears in the margin near the relevant narrative passage.

1. *John Ruwwe, Tommie Futch, and Ted Fullerton at the entrance of Spider Cave. This cave near Carlsbad Caverns was the first "wild" cave explored by the author and the site where the caving bug first bit him. Page 10. (July 16, 1933)*

1

Carlsbad, Caves, and a Camera   199

2

3

4

200   *Carlsbad, Caves, and a Camera*

2. *John Ruwwe, Ted Fullerton, and Dave Wilson in the Snow Room of Spider Cave. Page 17.* (July 16, 1933)
3. *Tommie Futch, Ted Fullerton, and Dave Wilson and the Tiger's Mouth in Spider Cave. Page 17.* (July 23, 1933)
4. *The first of the helictite clusters we came to call Medusa Heads in Spider Cave. Page 17.* (April 14, 1934)
5. *Medusa Heads in Spider Cave. Page 18.* (April 14, 1934)
6. *Helictites and soda straws in Spider Cave. Page 18.* (April 14, 1934)
7. *Dave Wilson, Fred Burgess, and Cal Livingston near the Room of the White Pillars, Spider Cave. Page 19.* (April 14, 1934)

6

8. *The Fairy's Wand, a soda-straw stalactite tipped with a cluster of aragonite crystals, hanging in a passage in Spider Cave. Pages 20, 23. (April 14, 1934)*

9. *Dave Wilson beside Cactus Spring deep in Spider Cave. Tiny aragonite crystals, sharp as cactus spines, covered the rim of the pool and grew up the sides of the formations, giving the pool its name. Page 20. (April 14, 1934)*

10. *The two Canadians, Fred Burgess and Cal Livingston, in the Room of the White Pillars in Spider Cave. Page 23. (April 14, 1934)*

8

11

12

11. *Hershel Davis, C. W. "Sonny" Hagler, Roy "Sparky" Renfro, Bacel "Scotty" Scott and the author against a background of Medusa Heads in Spider Cave. Pages 28, 30. (Summer, 1935)*
12. *Chocolate pagodas and a white temple deep in Spider Cave. Page 30. (December 30, 1934)*
13. *The Chocolate Sundae almost fills a secluded room in an upper section of Spider Cave. Page 29. (Summer, 1935)*

14

15

16

14. *The old wooden stairway down through the entrance of Carlsbad Caverns. This replaced the original primitive guano bucket hoist, and was in turn replaced by the inclined trails now in use. Page 33. (C. 1925; photo by Ray V. Davis)*

15, 16. *Early Davis photographs in the Hall of the Giants in Carlsbad Caverns. Jim White, original Caverns explorer, is at upper right in white sombrero. Page 34. (C. 1920)*

17. *Another early Davis photograph at the Totem Poles, Big Room, Carlsbad Caverns. Page 34. (C. 1922)*

17

18, 19. *Early Davis photographs of the Totem Pole area in Carlsbad Caverns. Note early place-names, long since replaced, in Photo 18. Jim White is seated on stalagmite at right. In Photo 19, Jim White in white shirt and J. R. Yates lean on stalagmite in center. A. W. "Pete" Anderson is at right, looking up. Ray had begun realizing some returns from his efforts, both in publicity and in sales of his photographs. Note the copyright No. 1. Page 34. (C. 1920)*
20. *One of Ray Davis's best known early photos in Carlsbad Caverns—the Twin Domes. Jim White sits at right. Page 34. (C. 1920)*

21. *The author's first photo at Carlsbad Caverns, looking out the great entrance. Page 35. (August 20, 1926)*

22. *The author's first photo inside Carlsbad Caverns—a time exposure in the King's Palace. Page 35. (June 2, 1927)*

23. *The author's small camera, placed at one side, caught him and Eugene Davis and the big 8 × 10 camera as Ray Davis lighted a scene to the left. The Giant Dome in Big Room, Carlsbad Caverns, makes the background. Page 36. (January 15, 1928; photo by Ray V. Davis)*

22

24

25

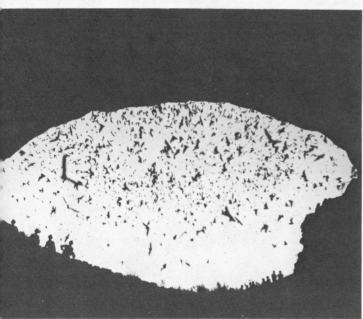

24, 25, 26. *Carlsbad Caverns "goes Hollywood"! Stills from filming* The Medicine Man *show leads Tom Santchi and Jean Layman (in blackface) watch villain Philo McCullough go crazy and die with his ill-gotten treasure. Page 39.* (June 28, 1929)

27. *Evening bat flight from the entrance of Carlsbad Caverns. Page 40.* (Summer, 1935)

27

Carlsbad, Caves, and a Camera   213

Carlsbad, Caves, and a Camera

28. *The graceful Temple of the Sun in Big Room, Carlsbad Caverns. This is one of the most popular spots with the thousands of tourists who yearly make the underground trek. Page 40.* (Summer, 1935)

29. *First Carlsbad Caverns explorer, Jim White, and the author. It was Jim who directed the author to his first "wild" cave. Page 9.* (Spring, 1933)

Carlsbad, Caves, and a Camera   215

216　　*Carlsbad, Caves, and a Camera*

30. *Tommie Futch, Julian Shattuck, the author, Dave Wilson and Ted Fullerton at the entrance of Hell Below. Page 53.* (September 10, 1933)

31. *Headed home—Ted Fullerton, Tommie Futch, Dave Wilson and Julian Shattuck as the sun sinks beneath the ridge above Hell Below. Page 54.* (September 10, 1933)

32. *Looking up the rope that marooned the cave explorers in Hell Below. Page 58.* (October 1, 1933)

32

34

33. *Dave Wilson at the entrance of Falling Rock Cave, where two explorers met disaster. Page 61.* (February 3, 1934)

34. *The author, bloody but unbowed, after an accident in Falling Rock Cave. Page 63.* (February 3, 1934)

35. *Julian Shattuck after Falling Rock Cave struck a second time. Page 64.* (February 3, 1934)

35

220 Carlsbad, Caves, and a Camera

36. *Early photographic party camped near Hidden Cave. Carl Livingston, pioneer writer and explorer, is in the white shirt; George Pixler sits just in front of the car. Page 65.* (C. 1926; photo by Ray V. Davis)

37. *The author in the locked barricade built by the Forest Service over the entrance of Hidden Cave. This was an early attempt to prevent vandalism of the cave. Page 66.* (February 4, 1934)

38. *The War Club Room in Hidden Cave. Page 67.* (February 4, 1934)

39

40

39, 40. *Carl Livingston (upper left photo); and Seth McCollum (bottom), over a decade later, lost in a section we called Fairyland in Hidden Cave. Page 67. (Photo 39, c. 1926, by Ray V. Davis; photo 40, February 4, 1934)*
41. *Seth McCollum by the Leaning Tower in Hidden Cave. Page 68. (February 4, 1934)*
42. *Dave Wilson in the Coral Garden of Hidden Cave. Page 68. (February 4, 1934)*

43. *Julian Shattuck in the Icicle Room of Hidden Cave, where every stalactite bore on its tip a drop of glistening water. Page 68.* (February 4, 1934)

Carlsbad, Caves, and a Camera

44. *Dave Wilson, Seth McCollum and Julian Shattuck at the difficult and unused second entrance into Hidden Cave. Page 68.* (February 4, 1934)

45. *The Chinese Wall in Hidden Cave. Carl Livingston leans against the brown teepee in the middle distance. Page 68.* (C. 1926; photo by Ray V. Davis)

Carlsbad, Caves, and a Camera

46. *Carl Livingston (squatting center), Pete Anderson (pipe in mouth), and Huling Ussery (far distance) survey the beauties of Hidden Cave. Ray Davis got his behind in this picture, as he squatted, center foreground, to blow a flare from his magnesium gun. Page 69.* (C. 1926; photo by Ray V. Davis)

47. *A room of rare beauty in Hidden Cave. Page 69.* (C. 1926; photo by Ray V. Davis)

48. *The Coral-Room-Looking-For-A-Disaster in Hidden Cave. Page 69.* (February 4, 1934)

48

49. *Julian Shattuck amid the great coral heads and sponge-like formations of the Lost Chamber of Hidden Cave. Page 70. (February 4, 1934)*

50. *Julian Shattuck, Dave Wilson, and Seth McCollum take a lunch break in the Fairies' Palace of Hidden Cave. Page 71. (February 4, 1934)*

51. *The author in the Grand Canyon of Hidden Cave. Page 72. (February 4, 1934)*

52

53

52. *Tommie Futch, the author, Glenn Hamblen, and Dave Wilson at the entrance of Endless Cave. Page 75. (March 4, 1934)*

53. *A forest of white stalactites decorates the ceiling of the entrance passage of Endless Cave. Page 76. (June 8, 1937)*

54. *Sonnie Kindel, Dave Wilson, Glenn Hamblen, and Tommie Futch pause for a breather at First Spring in Endless Cave. Page 77. (March 4, 1934)*

55. *Sonnie Kindel on the high balcony and Charlie Stevens by the Totem Pole in the Grand Canyon of Endless Cave. Page 80. (June 8, 1937)*

54

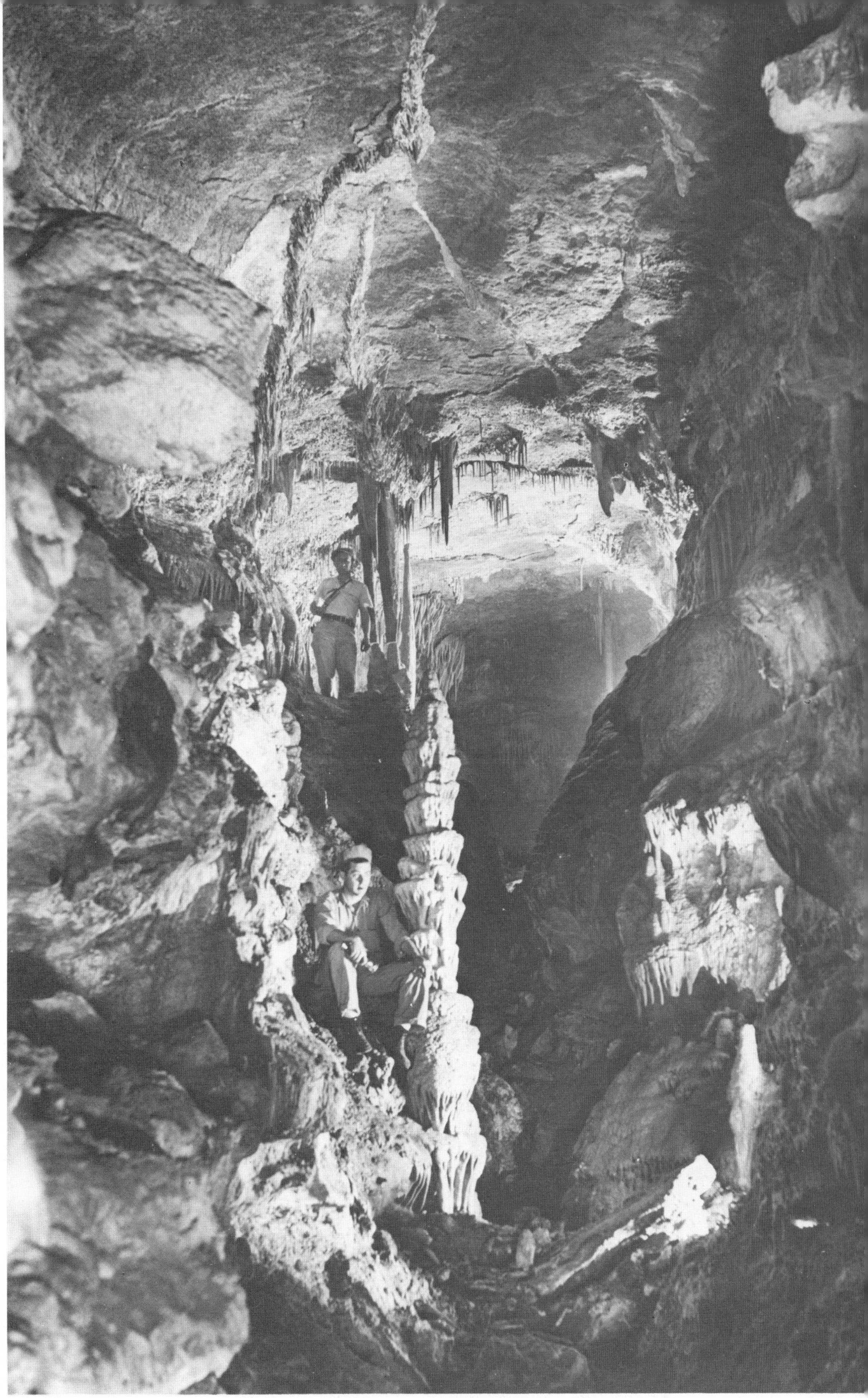

56. *Dave Wilson at the top, Fred Burgess, Cal Livingston in the Whale's Mouth, and the author at Taffy Hill of Endless Cave. Page 80. (April 11, 1934)*
57. *Fred Burgess at Lily Pad Spring in Endless Cave. Page 81. (April 11, 1934)*
58. *Dave Wilson, the author, and Ray Sims in The Distillery of Endless Cave. Page 83. (March 11, 1934)*
59. *Ray Sims and Dave Wilson in the Soda Straw Room of Endless Cave. Page 84. (March 11, 1934)*

Carlsbad, Caves, and a Camera 235

Carlsbad, Caves, and a Camera

60. *The beautiful Soda Straw Room in Endless Cave. The white straw hanging at the left was over ten feet long. Page 84. (March 25, 1934)*
61. *Marge Dwyer, the author's guest from Silex, Missouri, in the stone draperies of the Soda Straw Room in Endless Cave. Pages 84, 85. (August 26, 1936)*

62. *Ray Sims and Dave Wilson in tiny Captain Nemo's Grotto of Endless Cave. Page 84. (March 11, 1934)*
63. *The author, Sonnie Kindel, Charlie Stevens, and Tommie Futch pause for a break in the Badlands of Endless Cave. Page 85. (June 8, 1937)*
64. *Stately Columns and stone draperies in Endless Cave. Page 85. (June 8, 1937)*
65. *Mary Nell Berger and Sonnie Hagler in the Coral Room of Endless Cave. Page 86. (August 20, 1936)*

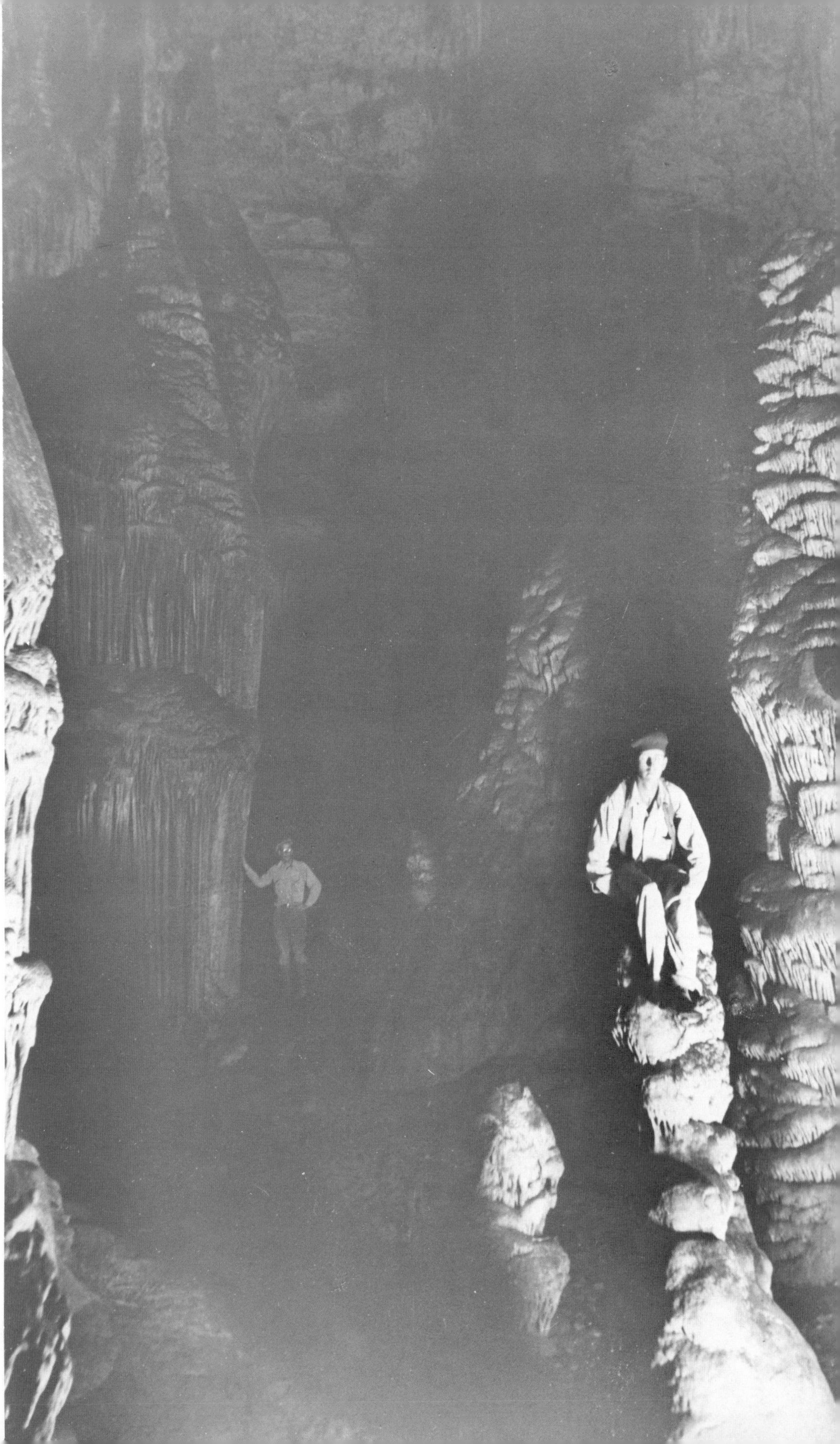

67

66. *In the vast entrance hall of Cottonwood Cave. Julian Shattuck perches atop a stalagmite, while Ted Fullerton leans against a giant column at the right. Dave Wilson is almost lost in the far darkness amid standing monoliths. Page 90.* (September 10, 1933)

67, 68. *Dave Wilson and Ted Fullerton seem like dwarfs at the base of the Goliath, one of the world's largest stalagmites, in Cottonwood Cave. The lower photo shows only the upper one-fourth of this massive formation. Page 92.* (September 10, 1933)

69

70

Carlsbad, Caves, and a Camera

69. *The high country of Slaughter Canyon, one of the most productive caving regions in the Guadalupes. Page 95.* (January 7, 1934)
70. *The great arching entrance of Goat Cave. Page 96.* (January 7, 1934)
71. *Tommie Futch, Ted Fullerton, and Glenn Hamblen in the entrance arch of Goat Cave. Page 96.* (January 7, 1934)
72. *Glenn Hamblen in the Domed Room of Goat Cave. Page 97.* (January 7, 1934)

72

73

74

73. *Early-day explorers in Grey's Cave. The location of this cave is as much a mystery as the identity of the cavers pictured here. Page 98.* (C. 1900? Photographer unknown)

74. *On the way to Grey's Cave, the mysterious and lost cave of the Guadalupes. Page 98.* (C. 1900? Photographer unknown)

75. *Going caving the hard way! The old Model-A stuck on a gravel bank on the way to Gunsight Cave. Dave Wilson takes a breather from pushing, while the author at the wheel peers back to determine the loss of power. Page 98.* (April 22, 1934; photo by Fred Burgess)

76. *The great entrance of Gunsight Cave looms far above on the slopes of Gunsight Canyon. Page 99.* (April 22, 1934)

77. *The Black Pagoda stands in somber majesty high on a ledge in Gunsight Cave. Page 100. (March 19, 1939)*
78. *Cal Livingston, Fred Burgess, and Dave Wilson in the Giant Column Dome of Gunsight Cave. The author stands at bottom. Page 100. (April 22, 1934)*

79. *Bill Burnet in the shadowed left foreground and Seth McCollum on the skyline seem tiny toy figures beneath the soaring arch of Gunsight Cave. Page 101. (March 19, 1939)*
80. *"Overwhelming!" Bill Burnet called the awesome drop into darkness of the entrance of Gunsight Cave. Page 101. (March 19, 1939)*

Carlsbad, Caves, and a Camera   247

81. *Seth McCollum, on mound in center distance, as seen from the room of the Black Pagoda in Gunsight Cave. Page 101.* (March 19, 1939)

82. *In the entrance auditorium of Gunsight Cave, Seth McCollum ponders the situation, while Bill Burnet takes a siesta after digging for bones of prehistoric mammals. Page 101.* (March 19, 1939)

83. *Dave Wilson aids Tommie Futch in the descent of the entrance pit of Gyp Cave. Page 103.* (February 25, 1939)

84. *Dave Wilson makes the drop into Gyp Cave to ascertain the situation of Tommie Futch who has preceded him. Page 103.* (February 25, 1934)

84

85. *Sonnie Kindel tries the ladder dropping into the narrow cleft which opens into the pit of Chimney Cave. Page 105. (October 9, 1938)*
86. *Beautiful translucent elephant-ear draperies of a deep orange color in a side tunnel of Chimney Cave. Page 107. (October 9, 1938)*
87. *Sonnie Kindel headed out of Chimney Cave on the old ladder put there by Jim White many years previously. Page 108. (October 9, 1938)*

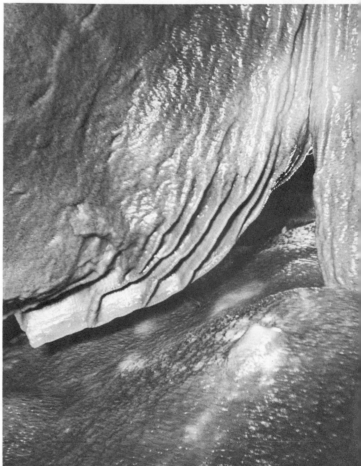

86

88. *Sonnie Kindel and Dave Wilson stand before the entrance of Dry Cave. Page 110.* (April 8, 1934)
89. *Glenn Hamblen checks out the crawlway leading into the cave he found in the entrance of Slaughter Canyon. We named it Hamblen Cave after its discoverer, unaware of the shadow of doom hanging over it. Page 115.* (April 29, 1934)

89

Carlsbad, Caves, and a Camera   253

90. *The gaping mouth of Kindel Cave, high in the wall of the Rattlesnake Bends of Dark Canyon. Page 119.* (August 8, 1937)
91. *Tommie Futch in the Black Temple Room of Kindel Cave. Page 122.* (August 8, 1937)
92. *Dave Wilson beside the Candy Waterfall in Kindel Cave. Page 122.* (January 28, 1934)

93. *Ted Fullerton beneath the elephant-ear draperies in Kindel Cave. Page 122. (January 28, 1934)*
94. *Ted Fullerton, Tommie Futch, and Dave Wilson look out of the entrance of Kindel Cave. The anchor bolts in the top of the arch, put there to aid in an early guano mining operation, are still in place. Page 123. (January 28, 1934)*

Carlsbad, Caves, and a Camera   257

95. *The Rhinoceros in Lake Cave. Page 128.* (June 23, 1937)
96. *Siamese Temple in Lake Cave. Page 128.* (June 23, 1937)
97. *Turreted wall of stalagmites in Lake Cave. Page 128.* (June 23, 1937)
98. *Tommie Futch by the Ice Cascade, with the Pope's Mitre in silhouette at the right in Lake Cave. Page 128.* (June 23, 1937)

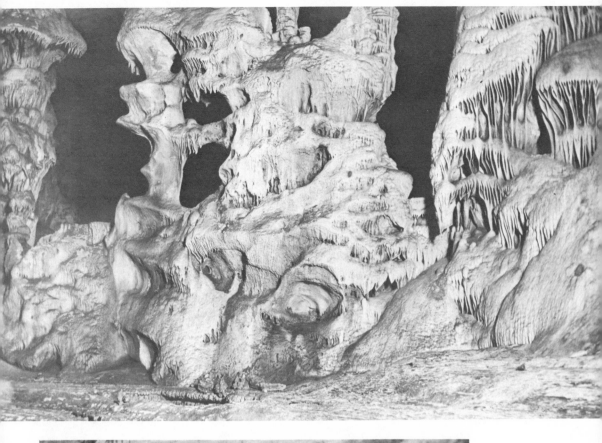

96

98

259

99. The mass of flowstone and stalagmites we called Castle Rock in Lake Cave. Page 128. (June 23, 1937)

100. The Backyard Gossip giving Tommie Futch an earful of the latest scandal in Lake Cave. Page 128. (June 23, 1937)

101. Like a shell-torn steeple in a war zone, the Twin Spires tower toward the ceiling in Lake Cave. Page 128. (June 23, 1937)

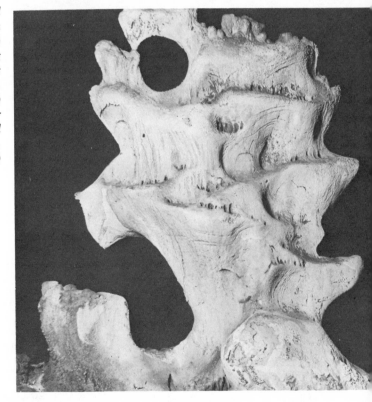

100

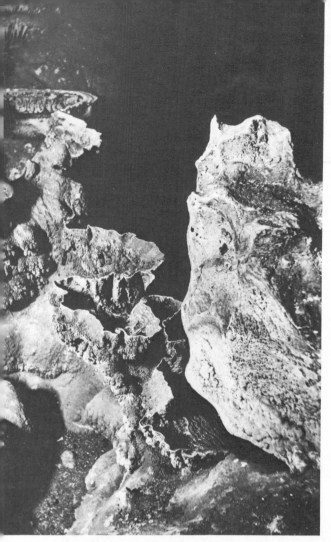

102. *A travertine dam forms the lip
of a precipice dropping off into the
Room of the Lake in Lake Cave.
Page 129.* (June 23, 1937)
103. *The Pointing Hand and the
Lighthouse rear up out of the cold
waters of the lake in Lake Cave.
Page 129.* (June 23, 1937)
104. *Color Me Blue! The author
takes the coldest swim of his life in
the lake in Lake Cave. Page 130.*
(June 23, 1937)
105. *A Niagara of stone covers a
section of wall in the Room of the
Lake in Lake Cave. Page 131.* (June
23, 1937)

106. *The entrance of Lodgepole Cave seems a remote spot on a distant landscape when viewed from the high ridge west of Gunsight Cave. Page 133. (March 19, 1939)*
107. *Tommie Futch watches as "Static" Burke tosses a rope into the larger of the two entrances into Lodgepole Cave. Page 134. (February 18, 1940)*
108. *Tommie Futch looks up the pine-tree ladder from the entrance of Lodgepole Cave. Page 134. (February 18, 1940)*
109. *"Static" Burke by the massive formation dominating the entrance chamber of Lodgepole Cave. The twin openings into the cave are at upper right and left. Page 135. (February 18, 1940)*

110. *Headed for a day's outing at McKittrick Cave. In left buggy are Robert Ezell and Frona Leck, while in the other are Bert and J. B. Leck, a niece of Bert's, and Deatron Campbell.* Page 136. (August 4, 1910; photographer unknown)

111. *Picnic at McKittrick Cave. Seated at left, clowning, is Bob Halley, and seated next to him is Zena Leck. Standing in profile is Gene Chapin, and at extreme right is C. D. Rickman.* Page 137. (C. 1898; photo probably by Bert Leck)

112. *In McKittrick Cave. In back center are Gene Chapin and Zena Leck. In front row are Bob Halley, Alice Leck, Maude Clark, two unknowns, C. D. Rickman, another unknown, and Bert Leck.* Page 137. (C. 1899; photo probably by Bert Leck)

113. *Exploring McKittrick Cave are an unknown man, then Frona Leck, Robert Ezell, Bert Leck's niece, Bert and J. B. Leck, and Deatron Campbell.* Page 137. (August 4, 1910; photographer unknown)

114. *Alice Leck, Bessie Berry, Zena Leck, Ola Kinchlow, and Maude Clark laugh it up in McKittrick Cave.* Page 137. (C. 1890; photo probably by Bert Leck)

115. *Bessie Berry and Bert Leck indulge in a favorite pastime—caving at McKittrick.* Page 137. (C. 1890 photo probably by Bert Leck)

111

112

116

117

118

268   *Carlsbad, Caves, and a Camera*

116. *Like a barbershop quartet, Harvey Hess, Frank Rainwater, Johnny Harvey, and Bert Leck strike a pose in McKittrick Cave. Page 137. (C. 1890; photo probably by Bert Leck)*

117. *Cavers sit for a formal portrait in McKittrick Cave. Three girls in front row are sisters Alice and Zena Leck and Maude Clark. At center back is Gene Chapin, and in front of him sits C. D. Rickman. Bert Leck is at extreme right in back row. Page 137. (C. 1890; photo probably by Bert Leck)*

118. *A new generation of cave explorers visit McKittrick Cave. Ruby Lee Jones, Grandpa Matthews, E. F. Brunneman, Forestina, Latina, and Lawrence Matthews, and Bertha Leck in back row, with Harold Renfro in front. Page 139. (March 3, 1929)*

119, 120, 121, 122. *The author, Tommie Futch, Glenn Hamblen, and Sonnie Kindel in the entrance to McKittrick Cave. Page 139. (March 4, 1934)*

121

123. *In the high country of Slaughter Canyon, from the north rim in the vicinity of Rainbow Cave. Page 139. (Summer 1935)*

124. *B. A. Nymeyer at the entrance of Rainbow Cave. Page 140. (March 15, 1931)*

125. *Tommie Futch at the entrance of Rainbow Cave, with the great precipices of Slaughter Canyon as a backdrop. Page 144. (January 14, 1934)*

123

124

125

126

127

126. *The Green Waterfall covering the right wall of the entrance chamber of Rainbow Cave. Page 141.* (March 15, 1931)

127. *The Orange Cascade and Great Column covering the left wall of the entrance auditorium of Rainbow Cave. Harold Renfro and B. A. Nymeyer stand between the two formations. Page 141.* (March 15, 1931)

128. *Dave Wilson stands beneath the canopy of the Canopied Green Throne in Rainbow Cave. Glenn Hamblen gave up and stopped about halfway up! Page 144.* (January 14, 1934)

129. *Right side of the Canopied Green Throne in Rainbow Cave. Page 142.* (January 14, 1834)
130. *The author leans against a great orange-colored stalagmite while brother B. A. climbs up on the Guardian in Rainbow Cave. Page 142.* (March 15, 1931)

130

131. *Glenn Hamblen and Tommie Futch at the base of Rainbow Cave's largest stalagmite, the Guardian. Page 142.* (January 14, 1934)
132. *The Guardian of Rainbow Cave stands silhouetted against the daylight pouring down through the great entrance. Page 142.* (March 15, 1931)

133. *The Painted Grotto in Slaughter Canyon, where Indian artists of the past covered the wall with myriads of pictographs. Page 147.* (February 18, 1934)

134. *Tommie Futch, "Dough" Spell, and Dave Wilson in the great, symmetrical entrance of Crystal Cave. Page 149.* (February 18, 1934)

135. *"Dough" Spell, Tommie Futch, and Dave Wilson in the frosty white beauty of the Snow Palace in Crystal Cave. Page 149.* (February 18, 1934)

134

136. *The Torture Chamber of Crystal Cave, where walls, floor, and ceiling are practically covered with hundreds of encrusted crystals of calcite, known as dog-tooth spar. Page 150. (February 18, 1934)*
137. *Sitting Bull Falls, an oasis in a desert canyon, behind whose falling waters is hidden an enchanting little cave. Page 151.*
138. *Nell Drury in the beautiful Drapery Room of Sitting Bull Falls Cave. Page 154. (August 25, 1946)*

137

Carlsbad, Caves, and a Camera   279

Carlsbad, Caves, and a Camera

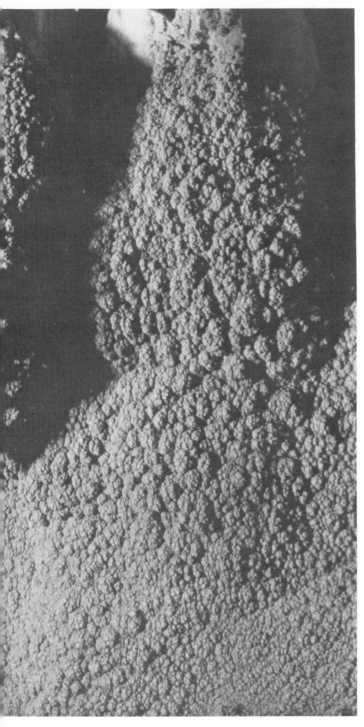

139. *Carl Livingston and the author in the chamber Carl called the Coral Room of Sitting Bull Falls Cave. This is one of the earliest—maybe the first—photographs made in this cave. Page 153. (April 19, 1931; photo by Carl Livingston)*

140. *The author's wife, Elizabeth, and son, Aaron, in Sitting Bull Falls Cave. The son lacked his father's enthusiasm for caving, and spent most of his time in the cave wanting out! Page 154. (August 25, 1946)*

141

141. *Tommie Futch and the author digging for a legendary treasure reportedly buried in Whistle Cave. Page 159. (Summer 1936)*

142. *Whistle Cave's only true stalagmite, slender and graceful, near the mysterious whistle. Page 160. (Summer, 1936)*

143. *Ancient Indian pictographs on the wall of Whistle Cave, just inside the entrance. Page 160. (Summer, 1936)*

144. *The ghost of the Indian artist threatening intruders into Whistle Cave? No, just the author making like a caveman silhouetted against the cave's entrance. Page 160. (Summer, 1936)*

142

143

144

145. *The entrance of Tunnel Cave high on the slopes of Slaughter Canyon, just as we first glimpsed it. Page 162. (December 30, 1936)*
146. *Tommie Futch by the stalagmites we called Mutt and Jeff on the entrance slopes of Tunnel Cave. Page 162. (December 30, 1936)*

Carlsbad, Caves, and a Camera

147. *Sonnie Kindel at the massive fluted column dominating the entrance chamber of Tunnel Cave. Page 162. (December 30, 1936)*
148. *Cactus-covered foothills of the Guadalupes, where we found Wind Cave. Page 164. (May 6, 1934)*
149. *Tommie Futch looks down into the mini-manhole entrance of Wind Cave. This kind of cave opening is a great deterrent to a fat man! Page 165. (May 6, 1934)*

149

150. *E. F. Brunneman in the entrance of New Cave, before gating. Page 170. (April 23, 1939)*
151. *E. F. Brunneman beneath the Terraced Waterfall in New Cave. Page 172. (April 23, 1939)*
152. *Charlie Stevens tests the stability of the Leaning Tower of Pisa in New Cave. Page 173. (November 27, 1938)*
153. *The Giant Toadstool in New Cave. Page 173. (September 18, 1938)*

152

153

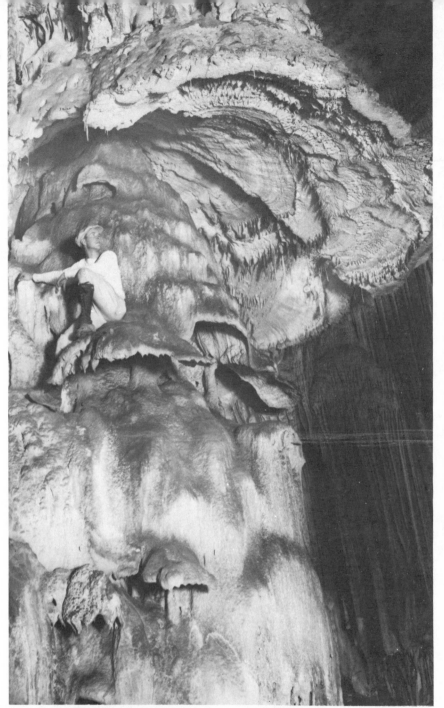

154

155

290

154. *Sonnie Kindel beneath the lip of the Giant Toadstool in New Cave. Page 173. (November 27, 1938)*
155. *The author in the Toad's Palace behind the Giant Toadstool in New Cave. Page 173. (September 24, 1938)*
156. *Mary Futch at the base of the Pillar of Hercules in New Cave. This is one of the most beautiful stalagmites in the cave. Page 174. (October 16, 1938)*

156

157. *Another view of the Pillar of Hercules in New Cave, with "Static" Burke for a size comparison. Page 175. (November 27, 1938)*

158, 159, 160. *These are all small, growing stalagmites; similar ones are found throughout New Cave. The first we called "Butterscotch Hills" because of the color. The second immediately became "Fried Ostrich Egg" for obvious reasons. The last appeared to be a "Sea Monster Surfacing." Page 175. (September 24, 1938)*

157

Carlsbad, Caves, and a Camera

161. *Another of the small, growing stalagmites in New Cave. This one we dubbed "Bristly Joe." Page 175. (October 16, 1938)*
162. *The Black Forest in New Cave. Page 176. (September 18, 1938)*
163. *Travertine dams, or dikes, in New Cave, seem to protect the Black Forest beyond from flash flooding. Page 176. (April 23, 1939)*

161

164. *The author seems a pigmy beneath the towering masses of flowstone, one of the predominant features of New Cave. Page 177.* (April 23, 1939)

165. *The author's mother, Zena, and Mary Futch amid the stalagmites clustered at the foot of the Giant Totem in New Cave. Page 177.* (October 16, 1938)

164

165

Carlsbad, Caves, and a Camera   297

166. *Sonnie Kindel beneath the canopy of the White Pagoda in New Cave. Page 177. (September 24, 1938)*

167. *The Chinese Wall, like a ribbon of sugar candy, winds its way across the level floor of New Cave. Page 177. (September 18, 1938)*

168. *Another view of the Chinese Wall in New Cave. Page 177. (September 18, 1938)*

168

169. *The author stands beneath the German Officer's Helmet in the fairyland region near the Christmas Tree Room in New Cave. Page 178. (November 27, 1938)*
170. *The Christmas Tree in New Cave, with Vince Lockhart dwarfed at its base. This is easily the most beautiful formation in the cave, with its glistening white mantel of calcite crystals. Page 178. (May 26, 1940)*

171. *Niobe's Tear hanging in an alcove of the Christmas Tree Room, like a teardrop of the weeping mother the Greek gods turned to stone. An impressive formation in New Cave. Page 179. (September 24, 1938)*

172. *The Clansman in New Cave, with the author crouching in front. This is one of the more awesome sights in the cave. Page 179. (September 24, 1938)*

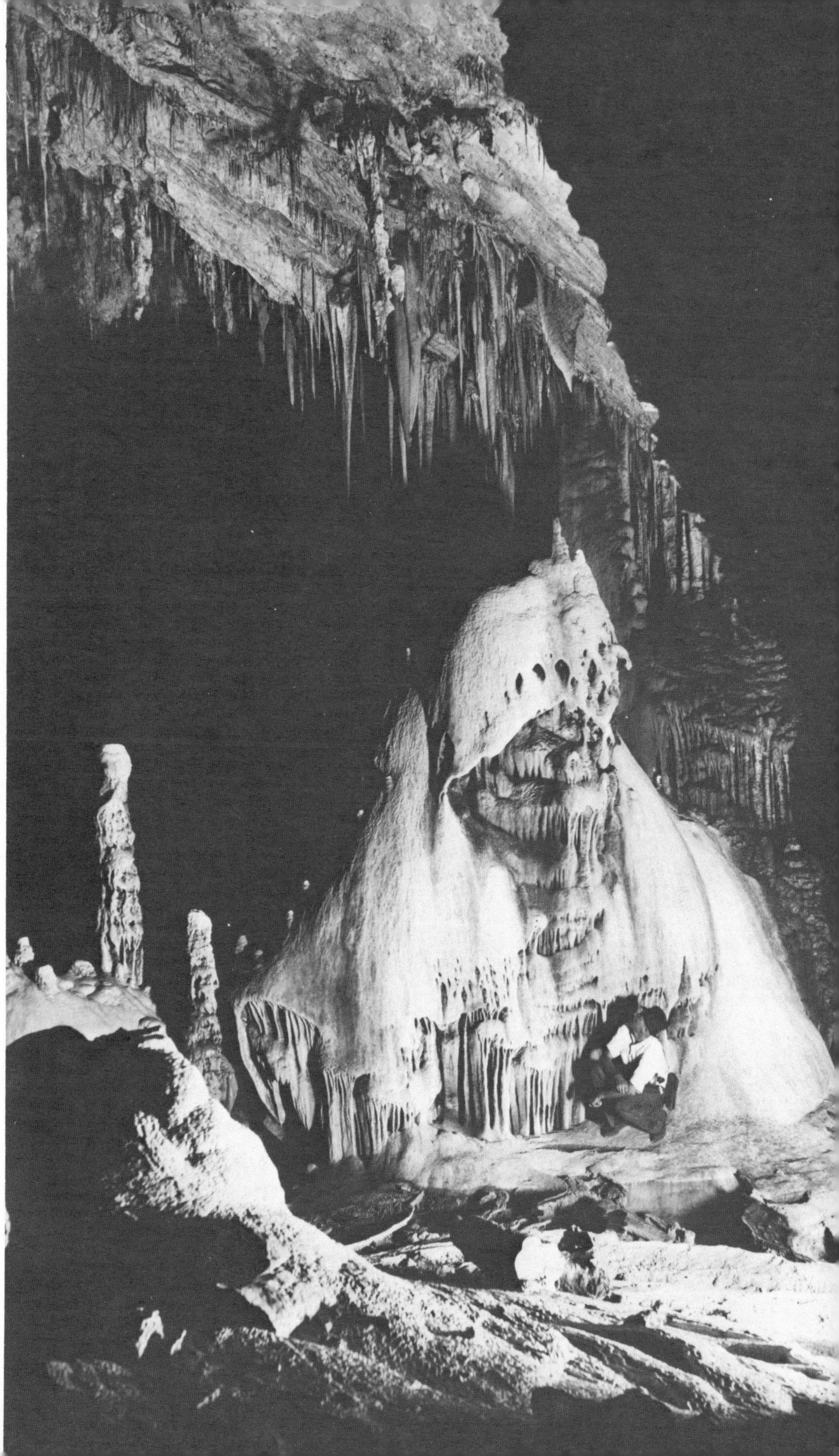

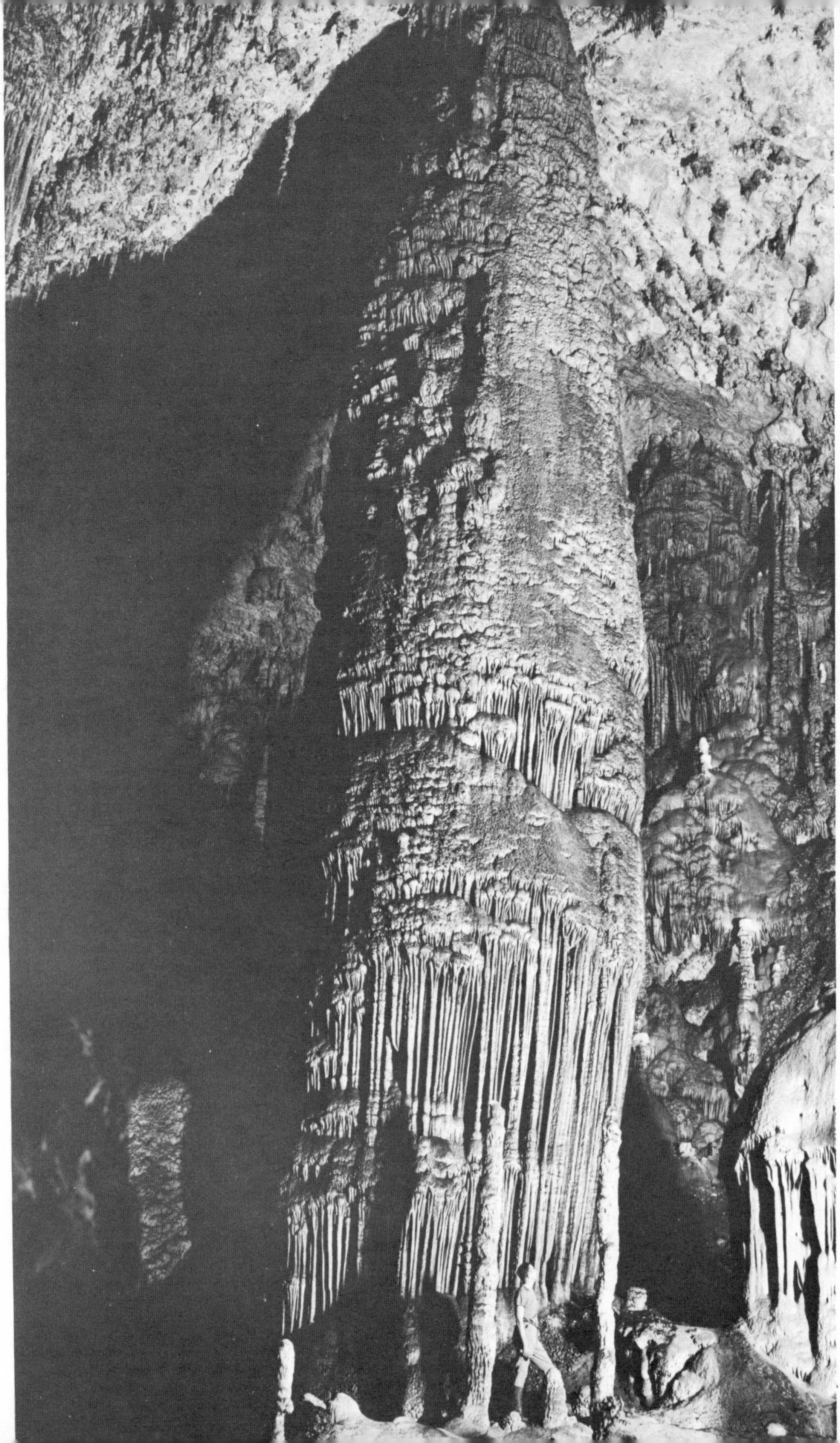

172

173. *Vincent Lockhart at the base of the Colossus, or Monarch, which is one of the world's largest stalagmites, in New Cave. Towering over 80 feet, it dwarfs all who stand beneath it. Page 180.* (May 26, 1940)
174. *Seen against daylight pouring in through the entrance of New Cave, the stalagmites on the opening slope seem eerie and unreal. Page 181.* (November 27, 1938)
175. *"Static" Burke seems lost in the maze of decorations in the White Palace of New Cave. Page 182.* (November 27, 1938)

175

Carlsbad, Caves, and a Camera

176. *The Guillotine hanging in the center of the White Palace of New Cave. Page 182. (September 24, 1938)*

177. *The Wreckage in New Cave, where disaster struck in ancient times, toppling stalagmites like tenpins. Page 183. (September 24, 1938)*

178. *The Fallen Giant in New Cave, with Mary Futch surveying the scene. Another proof of ancient earth tremors, this great stalagmite is now covered with many feet of massive flowstone. Page 184. (October 16, 1938)*

179. *Tommie Futch dips a refreshing drink from a hidden pool in a lost room beneath the Fallen Giant in New Cave. Page 187. (November 27, 1938)*
180. *The author searches for a route to the Mezzanine, past Old Baldy in New Cave. Page 190. (September 24, 1938)*

181. *Wayne Stell on his way to the Mezzanine, the high upper balcony in New Cave. Page 191.* (November 27, 1938)

182. *"Static" Burke at the Chinese Temple perched high on the Mezzanine in New Cave. Page 191* (November 27, 1938)

Carlsbad, Caves, and a Camera

183. *View from the Mezzanine in New Cave. Looking out over the Great Room, or Ballroom, one sees the Giant Totem dominating the center and the tiny figure at its feet. In the right distance gleams the canopied White Pagoda, or F. F. F. At extreme bottom center stands a small section of the Chinese Wall. The Christmas Tree Room and the Colossus are out of the picture to the left. The dude from the East is at the base of the Giant Totem, center, and Tommie Futch stands in left center in the bright flash. This photograph is a good example of the primitive multiple synchro flash method used. Four paper spoons of flash powder, as explained on pages 4 and 5, were lighted simultaneously; the dude lit the one in front of him to the right, Tommie lit the one in front of him, and Charlie Stevens lit the far one behind Tommie. The author at the camera lit one beside him to light foreground at right. Page 191.* (November 27, 1938)

184. *The author's brother-in-law, Bob Fleming, rests for a moment on the trail to New Cave. Behind him is the entrance of Slaughter Canyon, with the recent flood plain white in the distance.* Page 191. (October 4, 1942)

185. *The Guadalupe Mountains of New Mexico and Texas, where hundreds of caves, discovered and yet to be found, beckon the serious speleologist, the casual caver, and the novice adventurer in search of thrills and the unspoiled beauty of nature. Enough abounds here for all.* Page 191. (Summer, c. 1930)

184

185

Index of Persons